Job Demands in a Changing World of Work

A word cloud of the contents of the book

Christian Korunka · Bettina Kubicek
Editors

Job Demands in a Changing World of Work

Impact on Workers' Health and Performance and Implications for Research and Practice

Springer

Editors
Christian Korunka
Faculty of Psychology
University of Vienna
Vienna
Austria

Bettina Kubicek
Department of Applied Psychology
University of Vienna
Vienna
Austria

ISBN 978-3-319-54677-3 ISBN 978-3-319-54678-0 (eBook)
DOI 10.1007/978-3-319-54678-0

Library of Congress Control Number: 2017932784

Printed on acid-free paper

This Springer imprint is published by Springer Nature
The registered company is Springer International Publishing AG
The registered company address is: Gewerbestrasse 11, 6330 Cham, Switzerland

Contents

About the Editors

Christian Korunka is Professor of Work and Organizational Psychology at the Faculty of Psychology, University of Vienna. He has published journal articles and book chapters dealing with new job demands, organizational change processes, and quality of working life. His current research focuses on new forms of work (e.g., new office concepts, boundaryless work), new job demands and their role in shaping quality of working life. Another stream of his research deals with psychological aspects of entrepreneurship and family businesses.

Bettina Kubicek is Assistant Professor of Work and Organizational Psychology at the University of Vienna, Austria and Professor of Organizational Development at the University of Applied Sciences Upper Austria. Her current research focuses on new demands, especially on the positive and negative effects of work intensification and flexible and boundaryless work on employee well-being and the work-home interface. She is also interested in emotion work, older workers and retirement.

Chapter 1
Job Demands in a Changing World of Work

Christian Korunka and Bettina Kubicek

In recent decades, employees in the Western world have been affected by many changes in their working conditions. New types of work (e.g., crowd-working) and new work settings (e.g., flexible work arrangements) have appeared and employees at many workplaces seem to suffer from intensified work, which results in decreases in the quality of working life (Burchell et al. 2002; Green 2004). On the other hand, many improvements to working conditions are observable (Eurofound 2015). The world of work is getting more complex and study results are sometimes quite heterogeneous (Eurofound 2015). The aim of this book is to give an overview of changes and developments in the world of work over the last few decades.

Changes in the world of work have been triggered by fundamental societal and technological developments over the last few decades. Beginning in the 1980s in the United States and in the United Kingdom and further fostered by the collapse of the communist countries, there has been a strong development towards free market politics and reduced governmental control in many countries. "Neoliberalism" as the political foundation of free trade and globalization is leading to increased competition in practically all sectors of industry, production, trade and services. The economic pressure on production and services is thereby increasingly affecting many places of work. Longer working hours and reduced job security are just a few consequences on the employees' side.

These changes and developments are enabled and supported by broader developments in information and communication technologies (ICT). Wide use of computers in office workplaces started only in the 1980s. Since then, information and communication technologies have spread to practically all workplaces and also to virtually every aspect of our non-work lives. The rapid development of the

C. Korunka (✉) · B. Kubicek
Faculty of Psychology, Universitätsstraße 7, 1010 Vienna, Austria
e-mail: christian.korunka@univie.ac.at

B. Kubicek
e-mail: bettina.kubicek@univie.ac.at

C. Korunka and B. Kubicek (eds.), *Job Demands in a Changing World of Work*,
DOI 10.1007/978-3-319-54678-0_1

internet and of smartphone technologies further intensified the diffusion of ICT in all parts of society. New forms of communication technologies complemented personal communication structures in many ways.

All of these changes and developments seem to be occurring at an increased speed. The sociologist Rosa (2013) coined the term "social acceleration" for the increased pace of changes in late capitalism. He differentiates between three aspects of social acceleration: Technological acceleration is the "speeding-up of intentional goal-directed processes of transport, communication, and production" (Rosa 2003: 6). Acceleration of social change describes the increasing speed of social structural changes, while acceleration of the pace of life describes the 'compression' of actions and experiences, with the aim of handling more actions within shorter timeframes. Social acceleration is widely observable in the world of work, where ICT as tools of technological acceleration support the speeding-up of communication, decision-making, and production; organizational change processes are executed at ever-shorter time intervals and the pace of life in many organizations seems faster than just a few years ago.

All these phenomena are leading to a reorganization of work in many aspects. For instance, management by objectives has resulted in a work organization based on individual goals, often without the necessary guidelines and resources to reach these goals. Another change in work reorganization is the trend towards "boundaryless" work settings: It may no longer be necessary to work nine to five in a conventional work location. Instead, with the appropriate ICT support, many kinds of work can be performed anywhere with no time constraints. This type of work results in a huge increase in flexibility, leading to a wide range of potentially positive and negative outcomes.

In our own studies, we described and empirically investigated three types of new demands, which are all outcomes of changes in work and work organization. "Work intensification" describes a process characterized by an intensification of work effort and is needed to reach ever-higher work goals. "Intensified planning and decision-making demands" are observed in some circumstances when autonomy at work increases strongly. This may be the case when work goals are only roughly agreed upon or in boundaryless work settings. Finally, "intensified learning demands" are observable when there is an increasing need to adapt to organizational change processes and to learn to use new forms of information and communication technologies. All three demands have huge potential for both positive and negative outcomes, depending on work settings, resources, and individual needs.

This book aims to give an empirically based overview of such trends and developments in the current world of work. Parts of the book are based on a comprehensive research project which was funded by the Austrian Science Fund and realized between 2011 and 2016. In a series of qualitative and quantitative studies, the role of new job demands in the current world of work was investigated. The studies ranged from diary approaches focusing on daily work processes and longitudinal studies with a time frame of more than two years to biographical interviews.

The seven chapters of the book cover many topics related to new demands at work, ranging from an analysis of socioeconomic changes as background drivers of the new demands to practical recommendations for job design.

In Chap. 2, Jörg Flecker, Theresia Fibich and Klaus Kraemer describe socioeconomic changes and the reorganization of work from a sociological perspective. They identify three important trends: (1) the trend towards "financialization" in the Western world, which is strongly related to but also goes beyond neoliberal developments in that the radical liberalization of financial markets has affected economies and thereby all organizations and companies; (2) globalization and the development of international networks with their far-reaching effects, like the redistribution of work into other countries; and (3) informatization and digitalization in their role not only as drivers of these developments, but also as drivers of countless organizational changes and developments. The subjectification of work and boundaryless work are then described as the most important aspects of the reorganization of work as a direct effect of the socio-economic changes. This chapter sets the stage for the following chapters, which focus on the new job demands resulting from these socioeconomic transformations and job design recommendations.

In Chap. 3, Matea Paškvan and Bettina Kubicek describe the intensification of work, which has been empirically confirmed by many scholars over the last few decades (e.g., Burchell et al. 2002; Green 2004). The authors highlight the differences between work intensity, indicated for instance by longer working hours (work extensification) and work intensification, which is characterized by the need to work faster and conduct different work tasks at the same time. Based on the results of a series of empirical studies, work intensification is identified as a unique job demand which can be clearly separated from time pressure. A number of strategies on both the organizational and individual levels are discussed to prevent the potential negative effects of work intensification as a hindrance stressor.

Chapter 4 by Bettina Kubicek, Matea Paškvan and Johanna Bunner deals with the bright and dark sides of job autonomy. In its approach to and understanding of new job demands, this chapter can be seen as a core chapter of the current book. The "bright" side of job autonomy, as an important resource in many classic models of job demands and job resources, is contrasted with the "dark" side of autonomy, i.e. the fact that too much autonomy (an inherent feature of many new work contexts, like boundaryless work settings) can impair employee well-being. In the latter setting, it has been shown empirically that there is a curvilinear relationship between the amount of perceived autonomy in a job and job outcomes related to quality of working life. Warr's (1994) vitamin model, which describes the relations between features of work and quality of working life using a vitamin metaphor (some of which are positive up to a certain amount and may have even negative effects on health when this amount is exceeded), is used as an excellent theoretical base to explain these relations. The chapter also points to the fact that the positive effects of autonomy are contingent on the form of autonomy, e.g., method autonomy versus worktime/workplace autonomy.

Chapter 5 shifts the perspective from the job to the career level. Irina Nalis describes agency and meaning as resources for dealing with new career demands. She extends the focus on job-related planning and decision-making demands to planning and decision-making demands at the career level. The traditional career contract, combining job security, a long-lasting perspective, and well-defined opportunities for promotion has been largely replaced in the current world of work by an increasing need to self-directedly manage one's career. The "protean career" and "career crafting" are then discussed as examples of career models that are playing an increasingly important role in the current world of work. The results of a larger qualitative study on career narratives are presented, vividly illustrating the characteristics of career paths in a world of work driven by socio-economic changes and demands for increased learning and mobility.

In Chap. 6, Cornelia Gerdenitsch describes demands related to "new ways of working". This term is often used as an umbrella term for work characterized by geographic and temporal flexibility and that is supported by information and communication technologiesCommunication technologies. Three aspects of new ways of working, "flextime", "flexplace" and "nomadic work" are analyzed in terms of their potential to fulfill psychological needs (autonomy, competency, social relatedness, and structure). Based on a literature review, Gerdenitsch concludes that all three forms of flexible work arrangements help satisfy employees' need for autonomy and competence. With regard to the need for relatedness, she stresses that positive effects are likely to occur if regular physical or virtual encounters are conducted. Computer-mediated communication technologies might help in this regard. Finally, the need for structure (which may be in contrast to the three basic needs, and might also differ between employees) may be thwarted in flexible work arrangements, but employees with a high need for structure still have the freedom to create their own structure.

In Chap. 7, Roman Prem brings in another perspective on new demands at work when he analyzes the effects of a changing world of work on daily working life. He reflects on the role of within-person processes at work in relation to new demands. Thus, the focus here is on mechanisms linking new demands to quality of working life indicators. The importance of subjective appraisal and self-control in new work settings is described. This chapter once again emphasizes the critical role of need satisfaction as an evaluation criterion for new work settings.

Chapter 8, written by Christian Korunka, discusses challenges for job design in the current world of work. The important role of information and communication technologies, which have many positive effects on working conditions but are at the same time important drivers of intensification processes and work interruptions, is discussed. While the importance of classic job design approaches has not changed and many aspects (like workplace safety, ergonomic design, etc.) are just as important as before, for some job aspects, the design goal has been somewhat altered. For instance, instead of maximizing autonomy, the goal should be to optimize autonomy, always depending on the specific characteristics of a given type of job.

Acknowledgments We would like to thank everyone who has contributed to this book. First of all, we would like to express our gratitude to the authors for their work and the preparation of their chapters. Many contributions to this book are the result of studies realized in a large research project funded by the Austrian Science Fund (FWF 23377-G17). We would like to thank them for their financial support of this project. Finally, we would like to thank Johanna Ganslmeier, Elisabeth Dorfinger and Springer publishers for their valuable support in finalizing this book.

References

Burchell, B., Ladipo, D., & Wilkinson, F. (Eds.). (2002). *Job insecurity and work intensification*. New York: Routledge.

Eurofound. (2015). *First findings: Sixth European working conditions survey*. Retrieved from http://www.eurofound.europa.eu/publications/resume/2015/working-conditions/first-findings-sixth-european-working-conditions-survey-resume

Green, F. (2004). Work intensification, discretion, and the decline in well-being at work. *Eastern Economic Journal, 30,* 615–625.

Rosa, H. (2003). Social acceleration: Ethical and political consequences of a desynchronized high-speed society. *Constellations, 10,* 3–33.

Rosa, H. (2013). *Social acceleration. A new theory of modernity*. Columbia University Press.

Warr, P. B. (1994). A conceptual framework for the study of work and mental health. *Work & Stress, 8,* 84–97.

Author Biographies

Christian Korunka is Professor of Work and Organizational Psychology at the Faculty of Psychology, University of Vienna. He has published journal articles and book chapters dealing with new job demands, organizational change processes, and quality of working life. His current research focuses on new forms of work (e.g., new office concepts, boundaryless work), new job demands and their role in shaping quality of working life. Another stream of his research deals with psychological aspects of entrepreneurship and family businesses.

Bettina Kubicek is Assistant Professor of Work and Organizational Psychology at the University of Vienna, Austria and Professor of Organizational Development at the University of Applied Sciences Upper Austria. Her current research focuses on new demands, especially on the positive and negative effects of work intensification and flexible and boundaryless work on employee well-being and the work-home interface. She is also interested in emotion work, older workers and retirement.

Chapter 2
Socio-Economic Changes and the Reorganization of Work

Jörg Flecker, Theresa Fibich and Klaus Kraemer

2.1 Introduction

In the last three decades, work and employment have changed considerably in the European Union. Socio-economic changes such as financialization, the rise of the network economy and digitalization have given rise to new work requirements and employment conditions. The liberalization of financial markets on a global scale has had consequences far beyond the financial industry. Corporate governance in general has adapted to the new business environment even in continental European countries previously known for their 'patient capital' and interlinkages between manufacturing and the financial industry. Liberalization and the privatization of public services have also expanded the scope of capital investments and have brought hitherto sheltered fields of employment under the purview of financial capitalism. In many sectors, the new corporate governance fostered the cross-border relocation of production and the outsourcing of a range of business functions in order to lower costs and increase flexibility. This has resulted in new networked forms of economic activities, lengthening value chains and expanding production

J. Flecker (✉) · T. Fibich
Institute for Sociology, University of Vienna, Rooseveltplatz 2,
1090 Wien, Austria
e-mail: joerg.flecker@univie.ac.at

T. Fibich
e-mail: theresa.fibich@univie.ac.at

K. Kraemer
Institute for Sociology, University of Graz, Universitätsstraße 15/G4,
8010 Graz, Austria
e-mail: klaus.kraemer@uni-graz.at

C. Korunka and B. Kubicek (eds.), *Job Demands in a Changing World of Work*,
DOI 10.1007/978-3-319-54678-0_2

networks. Financialization and the external restructuring of businesses are, therefore, important socio-economic trends that need to be taken into account when analyzing developments at the workplace level.

In addition, the last few decades have been characterized by accelerated technological change. In particular, information and communication technologies have facilitated economic restructuring, the emergence of new business domains and far-reaching rationalization and organizational change at the workplace level. The digitization of work has not only meant changes in employers' skill needs, the declining relevance of some occupations and the emergence of new ones, it has also added a new dimension to the global labor market in the making.

Against the background of, and fostered by, these socio-economic trends, there has been a dynamic restructuring of work and employment. In large parts of the labor market, employment has become less secure and the low-wage sector has expanded in many countries. Together with high levels of unemployment, this has resulted in an increase in precarious living conditions. While precarity has always existed on the margins of society, it is now more widespread, expanding the zones of vulnerability and insecurity. Societal trends of increasing female labor market participation and changing household structures have had consequences for the relationship between gainful employment and private life. Together with human resource strategies, the use of enabling technologies and a general acceleration of social life, this has amounted to a trend of eroding boundaries of work and employment. Temporal flexibility, spatial mobility and new forms of ICT-based work have changed what is considered to be 'normal' work.

In part, these trends have contributed to changes in work that have resulted in increased autonomy for workers. Changes to work organization and the erosion of previous boundaries often lead to expanded flexibility and freedom, in particular for highly skilled workers. This is often referred to as the subjectification of work. However, this trend does not necessarily mean improved working conditions. Stricter indirect forms of control, more direct exposure to market forces and higher levels of overall insecurity can make enhanced levels of relative autonomy a mixed blessing. Moreover, there are strong countertendencies towards standardized forms of work organization leading to deskilling and a degradation of work in new environments. Seen against the background of various socio-economic trends, work organization is currently being shaped by opposite management strategies, often resulting in hybrid forms of standardized and subjectified work.

In this paper, we first address three central socio-economic developments, namely financialization, network economy and digitalization, which have prepared the ground for recent changes in work and employment. In the next section, we take a closer look at these changes, referring to academic debates about precarization, the blurring boundaries of work and contradictory dynamics of work organization. We conclude that future research needs to focus on how workers are able to deal with these new demands.

2.2 Socio-Economic Trends

2.2.1 Financialization

For decades, the financial industry and financial markets were largely overlooked by the social sciences. Financial markets played either a marginal role or no role at all in the debates on the transformation of modern capitalism and its consequences for labor, which were waged along the lines of organized and disorganized capitalism (Offe 1985) and Fordism and post-Fordism (Hirsch and Roth 1986). Only in recent years has the insight begun to take hold that investigating the transformation of modern capitalism requires taking a closer look at financial markets. This reorientation is reflected in the debate on 'financial capitalism' in particular (Windolf 2005, 2008), which – similar to the debate in comparative political economy and analyses of 'financialization' (Epstein 2005; Krippner 2011; Deeg 2012) – has addressed the increasing significance of the financial industry and financial markets in contemporary capitalism. The financialization debate extends beyond the discussions on the neoliberalization of economy and society in the 1990s, which focused on the rollback of state coordination and the expansion of market coordination as the new system of social order (Crouch and Streeck 1997). The trend observed here is not only the radical liberalization of financial markets and the emergence of a globally operating financial industry but also, and in particular, a significant shift in the appropriation of profits from manufacturing to the financial industry. This shift can also be analyzed in terms of a power shift from the entrepreneurial and manager class to the financial class that is explained by fundamental changes in the private ownership of the means of production. According to this analysis, the liberalization of financial markets and the emergence of a global financial industry since the 1990s have gradually surrendered control over the means of production to "new owners": institutional investors, private equity, and hedge funds. Historically, owner-managed enterprises were already seen being displaced by manager capitalism in the second half of the 20th century. What we are currently witnessing, it is argued, is that a type of ownership tied to an industrial base and to the means of production of an individual enterprise is being superseded by a new type oriented toward assets that are traded in capital markets.

The financialization thesis claims that businesses are increasingly aligned with the short-term profit expectations of global financial actors, and this observation is viewed in association with a fundamental structural transformation of modern Western-style capitalism. The emergence of a financially driven capitalism is identified at three levels: interests, ideas, and institutions. At the *first* level, a change in the constellation of economic interests is described that is considered to be rooted in altered ownership structures in enterprises. The *second* level refers to guiding principles or ideas. The diagnosis here is that we are witnessing a change in normative conceptions of what constitutes an efficient economic order. And *thirdly*, the emergence of a capitalism driven by financial markets is depicted at the political-institutional level as involving the deregulation of financial markets, financial products, and financial enterprises. The conclusion drawn is that this change in the configuration of interests, ideas, and institutions has come to pervade the different varieties of Anglo-Saxon and continental European capitalism alike (Hall and Soskice 2001; Jackson and Deeg 2006).

How is the emergence of financial capitalism explained in detail at each of these three levels of analysis? At the level of interests, a change in the ownership structures of enterprises is postulated, leading to an increasing dominance of institutional investors. These “new owners” (Windolf 2008) command an ever growing amount of capital that can be utilized for investments in global financial markets with the objective of reaping quick profits from gains. This investment strategy of ‘short-termism,’ the argument goes, has infused listed companies with a focus on short-term profits, with the result that organizational and management decisions get caught up in the maelstrom of the external logic of capital markets. These ‘new owners’, such as investment, hedge, and private equity funds, are considered less and less willing to share long-term entrepreneurial risks. They have the option of selling their shares in a globalized capital market at any point in time in order to diversify investment risks and seek more profitable investment opportunities in other sectors or world regions.

The increasing shareholder-value orientation and the focus on short-term profits have severely affected employment relations and management control in what was termed ‘disconnected capitalism’ (Thompson 2003): ‘Labour was asked to invest more of themselves (effort, commitment, new aspects of labor power such as emotions) at work, yet employers were retreating from investment in human capital, a retreat manifested in declining security, career ladders, pensions and the like’ (Thompson 2013: 473). Disconnected capitalism thus integrates breaches of social exchange and career contracts into the organization and the employment relationship (see also Nalis, Chap. 5). Organizationally, companies intensified ‘management-by-objective’ principles, leaving it to their decentralized units to reach set targets. ‘Business-within-a-business’ formations (Lacity et al. 2008) put pressure on company departments by setting return-on-investment targets, specified customer requirements, etc. (Marrs 2010: 344). This has resulted in a market-centered organization of work in which individual workers more directly than ever before face market requirements in their everyday working lives (Sauer 2010), intensification of work (see Paškvan and Kubicek, Chap. 3) and increased flexibility demands (see Prem, Chap. 7).

2.2.2 Networks Crossing Organizational and National Boundaries

Financialization as a particular form of corporate governance also has consequences for organizational structures. To increase the return on investment and boost shareholder value, various business functions and activities within them become subject to continuous make-or-buy decisions. Companies tend to concentrate on their core competencies and outsource other activities. These processes are often seen as a vertical disintegration or a hollowing-out of large corporations, resulting in a flexible network economy (Castells 2010b). Networks combine elements of hierarchies and markets, transcend corporate boundaries and cut across industry demarcations (Marchington et al. 2005; Windeler and Wirth 2010). In contrast to the accounts of a network economy, outsourcing can also be perceived as having resulted in the growth of new types of companies. In both the manufacturing and service sectors, large companies, in part with a global reach, offer to carry out production, IT or other business activities on behalf of clients in the private or public sectors (Flecker 2009). While it is true that the growth of these outsourcing providers has contributed to the decline of the vertically integrated manufacturing or service company, the networks that have emerged as a result do not necessarily consist of small or medium sized companies. In addition, what appears to be a move towards a network economy may simply be a restructuring of companies and industries. A case in point is the cooperation between brand owners (and partly hollowed-out corporations) such as Apple, Hewlett-Packard, Sony or Dell with the actually vertically integrated manufacturing company Foxconn (Lüthje 2006).

Depending on the industry and the occupations under investigation, different approaches to the analysis of economic structures and their consequences for labor seem appropriate. The Global Value Chain approach (Gereffi et al. 2005) focuses on inter-firm relations and, in particular, on the relationship between two links in a chain. This perspective highlights the dependency relations between a buyer company and its suppliers. This may reveal movements involving the economic upgrading or downgrading of companies and regions, resulting in improved or worsened opportunities for employment and job quality. However, social upgrading is not a necessary consequence of economic upgrading in the sense of 'the improvement in terms, conditions and remuneration of employment and respect for workers' rights, as embodied in the concept of decent work' (Barrientos et al. 2011: 301).

The existence and dynamics of value chains and production networks have far-reaching consequences for the analysis of working conditions. Whatever the direction of change in terms of upgrading or downgrading, the mere dynamic of value chain or production network restructuring may result in higher levels of insecurity for workers. The option of outsourcing (and re-outsourcing) and the continuous reshuffling of inter-firm relations imply a higher level of job insecurity as compared to the vertically integrated company which also relates to career prospects (see Nalis, Chap. 5). Inter-firm relations may also have an impact on other aspects of job quality. Depending on the relative power and contractual relations between the organizations, the client companies' strategies can affect pay, working hours, workload, flexibility demands, etc. to a greater or lesser degree. Service Level Agreements (SLAs), i.e. contracts between the client company and outsourcing provider, are an example of the influence the outsourcing company may have on immediate working conditions (Flecker and Meil 2010; Taylor 2010). Frade and Darmon (2005) argue that there is a general tendency within value chains to pass on risks and flexibility demands, leading to worse and worse working conditions further down the chain.

The dynamics of value chain restructuring and their spatial implications do have considerable consequences for the analysis of job quality at the company or establishment level: First, work may exist only temporarily, resulting not only in insecurity but also in the possibility of repeated far-reaching changes in work. Second, the position of the firm within (global) value chains and production networks matters. Dependency at the level of the inter-firm relationship has an effect on working conditions. Third, inter-firm relations may result, from the workers' perspective, in multi-employer constellations (Marchington et al. 2005), blurring the responsibilities of employers and complicating negotiations and interest representation. Fourth, outsourcing or vertical disintegration may result in spatial relocation. This may result in mobility demands on workers or in the need to cooperate, and often also simultaneously compete, over geographical distance and disparate time zones.

2.3 Informatization and Digitalization

The impacts of societal changes on work and employment are often closely related to the development of technology, its application and its usage. During the last few decades, many aspects of social change have been closely related to information and communication technologies because these have played a major role in societal spheres such as media and the arts, financial markets, politics and, not least, work. ICTs are based on, and simultaneously support, a long-term societal development called 'informatization', understood as the growing importance of information and information processing (Schmiede 1996, 2015). In history, double-entry bookkeeping was a prominent example; today it is Big Data. The expansion of information work as an enduring trend in capitalism has resulted in a convergence of many types of work in their outer form towards the use of computers, particularly in office environments. The technical integration of computers and communication technologies has intensified this trend, binding workers ever more tightly to their electronic devices and opening up further design leeway in the organization of work (see Kubicek, Chap. 4).

One aspect of socio-economic change is the acceleration of production and communication which, paradoxically, does not lead to more free time and leisure but rather goes hand in hand with an acceleration of the pace of life (Rosa 2005; Wajcman 2015). While complex societal and cultural developments have contributed to this outcome, it is obvious that people's sense of time has changed in an environment of instant communication. The internet and mobile ICT devices are often seen as contributing to information overload and the ensuing stress reactions. On the other hand, new information and communication technologies are also said to facilitate work-life balance (Pfeiffer 2012). Although research on the subject is not conclusive, observers seem to agree that the internet and mobile ICT devices can be seen as enabling factors contributing to the societal trend of blurred boundaries between work and private life (see below). The separation of information processing from material processes of production, transport or trade does have spatial implications. Not only can office buildings be separated from the factory floor, information work can also be relocated even further away. This is even more so the case if both the objects of work are immaterial or digital and the means of work consist of computer hardware and, in particular, software. 'Digital work' may be found in a range of industries and occupations from journalism, software development, and graphic design to tele-medicine. The shift to services, the tertiarization of manufacturing and the emergence of internet-related businesses have accelerated the spread of digital work. At the same time, these tendencies intensify the worldwide competition between educated workers, the global 'cybertariat' (Huws 2006).

While it is argued that digital work can be done anywhere, labor, as with all human activities, always 'takes place' in particular localities. These localities are far from being evenly distributed in geographic space. Rather, activities in the internet age are actually becoming more and more clustered, mainly in metropolitan areas, the 'nodal landing places' (Castells 2010a) of transnational networks. Thus, while remote and mobile digital work are the result of the tendency to free work from particular places, closer scrutiny reveals that place and distance retain their importance. Therefore, digital business and work in the current political economy do show particular historical 'spatial fixes' (Harvey 1982) even though the dynamics and fluidity may be much higher than in other industries or epochs. As a consequence, digital work is characterized by both 'placelessness' *and* stickiness, by both dispersion *and* spatial concentration (Flecker 2016).

Pervasive electronic networks, mobile internet applications and a wide variety of mobile devices not only allow for delocalized but also mobile work. In combination with organizational change and new forms of management control, these tendencies result in an accelerated erosion of the temporal and spatial boundaries between work and non-work spheres of life for a range of occupations. In part, digitization also calls the definition of work itself into question as more and more tasks are passed on to consumers or Internet users, who, though unpaid, are integrated into processes of value creation over the Internet (Ross 2013).

2.4 Contemporary Changes in Work and Employment

The socio-economic and socio-technological changes described above are embedded in and intertwined with developments in employment relations and work organization that have considerably altered the features of work and employment since the 1970s. These include the decline of mass-production manufacturing industries, the expansion of the service sector, the increase in educational levels, the privatization of public sector enterprises and public services, a decline of state regulation of employment and partly a shift of state responsibility for welfare provision to the market and to households (Watson 2008: 93). In the following section, we describe three central changes of work and employment shaped by these processes: precarization, blurring boundaries of work and contradictory dynamics of work organization.

2.4.1 Precarization

The term precarity does not describe just any kind of social change. Since it was first introduced in the work of Castel (2002), it has rather served to diagnose a profound transformation of contemporary work societies. Above all, it seeks to direct attention to the erosion of socially secure employment relationships. The main observation underlying this term is the return of types of wage labor (recommodification) that were believed to have been overcome under the conditions of Fordist capitalism and the 'Rhenish' welfare state. Stated more precisely, precarity characterizes an employment relationship that fails to meet certain social and legal standards that had hitherto been considered the norm and had been guaranteed by labor and collective bargaining laws as well as social policy and social security schemes. Gainful employment is precarious whenever the once common guarantees and legal rights are no longer fully granted or not granted at all.

Processes of precarization become apparent in negative deviations from the standards of normal employment or, more specifically, in structural discrimination compared to types of employment that correspond with the norm of the classical, socially protected 'standard employment relationship'. Seen from this angle, an employment relationship qualifies as precarious when employment and income are insecure in the long term, labor rights are restricted, and social status is fragile. A person in precarious employment is thus in a peculiar state of limbo (Kraemer 2010) in which there is constant hope that an insecure job might provide a springboard to stable employment, but this hope is accompanied by fears of social decline and of having to come to terms with precarious employment as a permanent state of affairs should all attempts at returning to secure employment fail.

One of the key findings of research on precarity (for a general account, see Castel and Dörre 2009; for Europe, see Mau et al. 2012; Scherschel et al. 2012; for the USA, see Kalleberg 2011) is that precarization is observed not only on the margins of society or among social strata that have always faced the threat of poverty but has spilled over into the middle classes over the past two decades. As a consequence, work research on the workplace level cannot take for granted the standard employment relationship and job security even at the core of the labor society and in high-skilled fields. This means that analytically, jobs and workers need to be located in particular labor market segments. In addition, job insecurity and status ambiguity need to be taken into account in the analysis of other dimensions of the quality of work in order to reach a reasonable understanding of working conditions and their consequences.

2.4.2 Boundaryless Work

In the 20th century, the predominant model of paid employment was characterized by a clear separation between the public and private spheres and often based on the model of the male breadwinner in standard employment and on female homemakers and caregivers. Since the 1970s, this organization has dissolved, profoundly changing the triad of standard employment, the welfare state and the nuclear family (Aulenbacher and Riegraf 2009). Increased female labor force participation, temporal flexibility, technological change and new management strategies have contributed to a blurring of the boundaries between paid work and workers' private lives and to the erosion of further institutions of work and employment. According to Voß (1998: 474), boundaryless work results from a social process 'of erosion and partly deliberate dissolution of social structures that regulate and delimit social processes and that have developed under particular historic conditions'. These shifting boundaries tend to restructure work and employment as a whole and are not only related to temporal and spatial aspects of work but also affect skills and competencies.

The blurring of *temporal* structures has progressed so far that today flexible working hours are perceived as 'normal' (Gottschall and Voß 2003: 17). In addition to the variety of forms of working hours (part-time, flextime, working time accounts, all-in contracts, on-call work) the beginning and the end of the workday, the location of working hours or work intensification periods are becoming increasingly situational and variable (Seifert 2007; Eichmann and Saupe 2014 for Austria). According to the sixth European Working Conditions Survey (EWCS), more than half of European employees work on Saturdays once in a month and one third work more than 10 h a day once in a month. In sum, 57% of European workers report having (highly) irregular working hours and arrangements; however, 43% do not (Eurofound 2015). This means that even the blurring of temporal boundaries does not affect all employees.

Regarding the *spatial* aspect, there are more and more virtual workplaces enabled by connections (internet, wifi and intranet), devices (laptops, smartphones and tablet computers), applications and services (email, calendars or access to databases and clouds) (Vartiainen 2006: 16). Work activities (even interactive work, Huws et al. 2004) can be carried out over spatial and temporal distance at home, on a train or even on a park bench (Halford 2005: 20). Aside from income or career aspirations, *personal identification* with work (Baethge 1991) serves as a motivation resource for why the boundaries between private and professional motivation have blurred (Voß 1998), preparing the ground for devotion and the self-exploitation of workers required to act as 'labor power entrepreneurs' (Voß and Pongratz 1998).

These new forms of working practices go along with new demands on employees (Voß 1998; Felstead 2005; Brodt and Verburg 2007). Although job autonomy is known as important job resource for employees, it can also become an obligation (Kubicek et al. Chap. 4). Workers are increasingly confronted with the need to self-organize their activities and to balance diverse and at times contradictory demands. For example, workers face the ambivalence of simultaneous self-control and stricter operational accountability or that of balancing the needs of family and work in highly boundaryless contexts (Heiden and Jürgens 2013: 40). In order to harmonize ambivalent requirements and highly contradictory demands (Heiden and Jürgens 2013: 38f), active border management is required for workers to maintain their own labor and 'life power' by defending, shifting, recreating or abandoning boundaries (Jürgens 2006).

A rather new example of boundaryless work is crowdworking: Crowdworkers perform tasks online which companies advertise on crowdsourcing platforms on the internet. 'More casual and short-lived forms of collaboration' replace conventional employer-employee relationships (Holts 2013: 40). Employees organize their time and workload and have to decide for themselves when and how long to work and which jobs to accept and which to decline. However, high competition and cost pressure depress remuneration, result in long working hours and require high availability on the platforms (Schörpf et al. 2016). This leads to 'client colonization' (Gold and Mustafa 2013) in the sense that workers' private lives are highly dependent on their clients. However, arrangements like teleworking or the opportunity to work from a home office also dissolve boundaries between home and work and result in changes to traditional forms of management control based on visibility and presence (Felstead et al. 2003). Furthermore, new office concepts such as hot desking, where employees are not assigned to desks and thus lose 'ownership of space' (Hirst 2011: 767), have been introduced, changing the way 'in which employees engage with the organization as well as who they are most likely to engage with' (Millward et al. 2007). Overall, these developments raise the question as to how demands for spatial and temporal flexibility allow for meeting psychological needs (see Gerdenitsch, Chap. 6).

2.4.3 Work Organization – Subjectification, Standardization and in Between

The emergence of the knowledge society is often assumed to have led to upskilling and higher levels of autonomy in work. Two main reasons are usually put forward: *First*, changing skill needs and new forms of management control have resulted in a more holistic use of labor power, in particular capitalizing on subjective capabilities such as ingenuity, creativity, responsibility, emotions, etc. This is in line with forms of management control characterized by greater autonomy for the workforce, which have been termed 'responsible autonomy' (Friedmann 1977), 'bureaucratic control' (Edwards 1981) or 'hegemonic control' (Burawoy 1979). These control mechanisms conceptualize workers to a lesser extent as unwilling subjects whose subjective influence on the labor process thus needs to be minimalized, as in Taylorism. In contrast, management acknowledges and demands intrinsic motivation, the search for self-fulfillment in work, workers' self-organization or devolved decision making. *Second*, people who have spent more time in the educational system and have attained educational and vocational qualifications seek to apply and further develop their skills, carry out demanding tasks in terms of content and participate in the shaping of work situations. This 'normative subjectivation' (Baethge 1991) of labor means that individuals make subjective demands on work and want to identify with it. Companies react to societal trends of individualization and educational expansion by increasing workers' discretion and opportunities for decision-making individually or in teams by introducing indirect forms of control such as management by objectives.

Subjectivity in work may be used either to compensate for the shortcomings of highly bureaucratic or Taylorized labor processes or to actively structure a less restrictive labor process (Kleemann et al. 2002). In their compensatory form, the subjective potentials of blue collar and white collar workers have always been used. However, in new forms of work organization this is no longer denied but workers are actively encouraged to bring in their subjectivity. As a consequence, in their structuring form, workers organize their own work and deploy their capacities to reach goals in an efficient way. This is why some authors argue that management control, i.e. making sure that labor power as a potential is actually being transformed into value-creating performance, is being shifted onto the working subjects themselves (Marrs 2010).

Subjectification is a particularly widespread tendency in the areas of high-skilled or creative work. Yet even there, autonomy may be limited to individual aspects of the labor process such as temporal or spatial ones, the content and skill dimension of work, the dimension of cooperation, emotional aspects, etc. (Lohr 2013: 431). And there are persistent tendencies to the contrary, i.e. the standardization of work. The prime example, of course, is Taylorism, characterized by distrust, deskilling and detailed surveillance (Braverman 1974). In recent years, research on call centers, public administration or shared service centers have provided examples of the standardization of work in the context of its informatization (Bain et al. 2002; Carter et al. 2013; Howcroft and Richardson 2012).

The restructuring and standardization of labor processes often goes along with the deskilling of workers. In order to achieve standardized procedures during the labor process, reliance on individual and personal engagement is reduced by way of the codification of tacit knowledge (Polanyi 1966), increasing the interchangeability of workers. The removal of individual expertise leads to a deskilling of workers, which in turn allows for lower wages (Howcroft and Richardson 2012). Howcroft and Richardson (2012: 120) analyzed the set-up of a shared service center and reported that management aims to achieve the 'right skill mix', that is to limit the number of skilled workers and workers with ambitious career plans. Therefore, promoted workers are replaced with the 'unofficially qualified' such as 'women returners and part-time students'.

The trend towards standardization alongside the one towards subjectification is being strengthened by processes of external company restructuring and the lengthening of value chains (Flecker and Meil 2010). The outsourcing of business functions are not 'lift and shift operations' (Ramioul and Van Hootegem 2015). Rather, outsourcing and relocation often require a far-reaching restructuring of the entire original labor process to be able to outsource locally or to relocate across borders (Howcroft and Richardson 2012; Ramioul and Van Hootegem 2015). This also means that 'placelessness', which is often ascribed to digital work as such, in reality needs to be actively produced (Flecker and Schönauer 2016).

By monitoring the process with standardized key performance indicators and other recording systems that streamline the production process (Ramioul and Van Hootegem 2015), management establishes new control mechanisms. This is true even for tasks that are not themselves standardized. Electronic monitoring and indicators allow for the graphical depiction of the performance and "achievement" of workers, sites and teams (Howcroft and Richardson 2012), providing immediate feedback on sales and other numbers. This puts increased pressure on workers, as these 'performance metrics' allow not only for surveillance from a distance by setting norms for productivity but also guide management's personnel decisions (on work intensification see Paškvan and Kubicek, Chap. 3). Therefore, 'the technology not only enables streamlining of processes, but also provides micro-level detail about their execution' (Howcroft and Richardson 2012: 118). However, standardization does not necessarily have to lead to a decrease in job quality, as bureaucratic control can increase labor process transparency and help in managing complex tasks (Adler 2005; Perlow 1999).

Although some scholars suggest there is a historical linear trend line of development from Taylorism towards subjectified forms of labor processes, research shows a coexistence of various forms of work organization in everyday working contexts (see for example Friedman 1977). In particular, recent contributions by labor process theorists refer to the use of hybrid forms of control (Sturdy et al. 2010; Reed 2010). Management combines various forms of control in order to 'juggle with the competing demands of market rationality, political expediency, social inclusivity, administrative conformity and legal integrity', giving a 'new solution to an old problem' (Reed 2010: 51). Prominent examples of such hybrid forms of control are industrial 'lean production', which combines fragmented and standardized work with team responsibility and continuous improvement processes, or 'Neo-Taylorism' in call centers (Sproll 2016) with standardized scripts, timelines and monitoring on the one hand and emotional work, which demands the mobilization of subjective capabilities, on the other. In elderly care as well, we find such hybrid forms of control between autonomy (individual interaction with clients according to the principles of good care) and standardization (narrowly clocked time slots and clear areas of operations) (Sørensen et al. 2015). Workers must meet the contradictory demands of each control form, and can hardly anticipate which type of management control will be applied in a concrete situation.

2.5 Summary

In the last few decades, we have observed changes in work and employment, which have speeded up since the beginning of the 21st century. In this paper, we have addressed three major socio-economic developments, namely financialization, the network economy and digitalization, that shape the contemporary organization of work, preparing the ground for the rise of phenomena such as precarization, boundaryless work and subjectified or standardized work organization.

The significance of financially driven capitalism has risen during the last few decades. Beside the liberalization of financial markets and the emergence of the globally operating financial industry, 'new owners' such as institutional investors, private equity and hedge funds have gained control over production. These 'new owners' are not willing to take on long-term risks, using their option of quickly selling shares on the global financial market in order to seek more profitable investments. In turn, businesses must orient themselves toward short-term profits in order to meet financial actors' expectations. Firms pass on these demands to their employees, introducing target values, decentralizing units, creating 'business within business' formations and simulating and confronting employees with market requirements within the firm. Management's reliance on workers' intrinsic motivation has led to the emergence of 'labor power entrepreneurs' (Voß and Pongratz 1998), which enhances workers' autonomy but also prepares the ground for workers' devotion and self-exploitation (see Kubicek et al., Chap. 4 on bright and dark sides of job autonomy).

However, there is not only a trend toward subjectified work forms for all workers. Vertical disintegration that keeps core competencies and outsources other activities has led to a flexible network economy and the emergence of global value chains. Relocation processes often require a massive restructuring of the entire labor process, resulting in the standardization and streamlining of work flows (Ramioul and Van Hootegem 2015). Yet even in highly standardized working environments, hybrid forms of work organization (Sturdy et al. 2010) combining subjectified and standardized forms of control can be observed.

As value chains are characterized by power and dependency relations between firms, firms pass flexibility demands down the value chain (Frade and Darmon 2005). The economic upgrading or downgrading of companies and regions can improve but also worsen conditions of work and employment like pay, working hours or workload. However, the dynamics of value chain or production network restructuring themselves result in higher levels of insecurity for workers. Multi-employer constellations (Marchington et al. 2005) complicate negotiation and weaken bargaining opportunities (e.g., threats of cross-border relocation).

As a third major driver of changing work organization, we referred to informatization and digitalization. On the macro level this trend goes along with a shift towards service work, the tertiarization of production and the emergence of internet business, resulting in new forms of work in the sense of information work, digital work or mobile work. On the employee level, electronic networks and mobile devices allow for constant accessibility, resulting in the blurring of former temporal and spatial boundaries between work and non-work spheres (Vartiainen 2006). Workers actively conduct boundary management in order to harmonize ambivalent requirements and highly contradictory demands and to obtain their own labor and life power (Jürgens 2006). In addition, other social structures which once regulated and delimited work are eroding, e.g., the 'standard employment relationship', as precarious working conditions and atypical employment are found not only at the margins of society, but have already reached the middle classes.

In conclusion, recent socio-economic changes have given rise to new forms of work, resulting in new job demands for employees. The dominance of financial markets, a global network economy and digitalization have facilitated greater autonomy and responsibilities for workers, but have also been used to standardize and deregulate work, leading to a rise in precarious employment and blurring the boundaries between work and non-work activities on various dimensions. Future research needs to address how workers can cope with these new requirements.

References

Adler, P. S. (2005). The evolving object of software development. *Organization, 12,* 401–435.

Aulenbacher, B., & Riegraf, B. (2009). Markteffizienz und Ungleichheit - Zwei Seiten einer Medaille? Klasse/Schicht, Geschlecht und Ethnie im Übergang zur postfordistischen Arbeitsgesellschaft. In B. Aulenbacher & A. Wetterer (Eds.), *Arbeit. Perspektiven und Diagnosen der Geschlechterforschung. Forum Frauen- und Geschlechterforschung*. Bd. 25 (pp. 230–248). Münster.

Baethge, M. (1991). Arbeit, Vergesellschaftung, Identität. Zur zunehmenden normativen Subjektivierung der Arbeit. *Soziale Welt, 42,* 6–19.

Bain, P., Watson, A., Mulvey, G., Taylor, P., & Gall, G. (2002). Taylorism, targets and the pursuit of quantity and quality by call centre management. *New Technology, Work and Employment, 17,* 170–185.

Barrientos, S., Gereffi, G., & Rossi, A. (2011). Economic and social upgrading in global production networks: A new paradigm for a changing world. *International Labour Review, 150,* 319–340.

Braverman, H. (1974). *Labor and monopoly capital: The degradation of work in the twentieth century*. New York [u.a.]: Monthly Review Press.

Brodt, T. L., & Verburg, R. M. (2007). Managing mobile work–insights from European practice. *New Technology, Work and Employment, 22,* 52–65.

Burawoy, M. (1979). *Manufacturing consent. Changes in the labor process under monopoly capitalism*. Chicago, London: University of Chicago Press.

Carter, B., Danford, A., Howcroft, D., Richardson, H., Smith, A., & Taylor, P. (2013). 'Stressed out of my box': Employee experience of lean working and occupational ill-health in clerical work in the UK public sector. *Work, Employment & Society, 27,* 747–767.

Castel, R. (2002). *From manual workers to wage laborers: Transformation of the social question*. New Brunswick: Transaction Publishers.

Castel, R., & Dörre, K. (Eds.). (2009). *Prekarität, Abstieg, Ausgrenzung. Die soziale Frage am Beginn des 21. Jahrhunderts*. Frankfurt am Main: Campus Verlag.

Castells, M. (2010a). Globalisation, networking, urbanisation: Reflections on the spatial dynamics of the information age. *Urban Studies, 47,* 2737–2745.

Castells, M. (2010b). *The rise of the network society*. Chichester: Wiley-Blackwell.

Crouch, C., & Streeck, W. (Eds.). (1997). *Political economy of modern capitalism. Mapping convergence and diversity*. London: Sage.

Deeg, R. (2012). Financialisation and models of capitalism. A comparison of the UK and Germany. In C. Lane & G. T. Wood (Eds.), *Capitalist diversity and diversity within capitalism* (pp. 121–149). London: Routledge.

Edwards, R. (1981). *Herrschaft im modernen Produktionsprozess*. Frankfurt am Main, New York: Campus.

Eichmann, H., & Saupe, B. (2014). *Überblick über Arbeitsbedingungen in Österreich. Follow-up-Studie. Sozialpolitische Studienreihe. Band 15*. Wien: BMASK.

Epstein, G. A. (Ed.). (2005). *Financialization and the world economy*. Cheltenham: Edward Elgar Publishing.

Eurofound. (2015). *First findings: Sixth European working conditions survey*. URL: https://www.eurofound.europa.eu/sites/default/files/ef_publication/field_ef_document/ef1568en.pdf. Accessed March 2016.

Felstead, A. (2005). *Changing places of work*. Basingstoke: Palgrave Macmillan.

Felstead, A., Jewson, N., & Walters, S. (2003). Managerial control of employees working at home. *British Journal of Industrial Relations, 41,* 241–264.

Flecker, J. (2009). Outsourcing, spatial relocation and the fragmentation of employment. *Competition & Change, 13,* 251–266.

Flecker, J. (Ed.). (2016). *Space, place and global digital work*. London: Palgrave Macmillan.

Flecker, J., & Meil, P. (2010). Organisational restructuring and emerging service value chains—implications for work and employment. *Work, Employment & Society, 24,* 1–19.

Flecker, J., & Schönauer, A. (2016). The production of 'placelessness'—digital service work in global value chains. In J. Flecker (Ed.), *Space, place and global digital work*. London: Palgrave Macmillan (forthcoming).

Frade, C., & Darmon, I. (2005). New modes of business organization and precarious employment: Towards a recommodification of labour? *Journal of European Social Policy, 15,* 21–107.

Friedman, A. (1977). *Industry and labour: Class struggle at work and monopoly capitalism*. London: Macmillan.

Gereffi, G., Humphrey, J., & Sturgeon, T. (2005). The governance of global value chains. *Review of International Political Economy, 12,* 78–104.

Gold, M., & Mustafa, M. (2013). 'Work always wins': Client colonisation, time management and the anxieties of connected freelancers. *New Technology, Work and Employment, 28,* 197–211.

Gottschall, K., & Voß, G. (Eds.). (2003). *Entgrenzung von Arbeit und Leben. Arbeit und Leben im Umbruch*. München [u.a.]: Hampp.

Halford, S. (2005). Hybrid workspace: Re-spatialisations of work, organisation and management. *New Technology, Work and Employment, 20,* 19–33.

Hall, P. A., & Soskice, D. (2001). *Varieties of capitalism: The institutional foundations of comparative advantage*. Oxford: Oxford University Press.

Harvey, D. (1982). *The limits to capital*. Oxford: Basil Blackwell.

Heiden, M., & Jürgens, K. (2013). *Kräftemessen. Betriebe und Beschäftige im Reproduktionskonflikt*. Berlin: Edition Sigma.

Hirsch, J., & Roth, R. (1986). *Das neue Gesicht des Kapitalismus. Vom Fordismus zum Postfordismus*. Hamburg: VSA.

Hirst, A. (2011). Settlers, vagrants and mutual indifference: Unintended consequences of hot-desking. *Journal of Organizational Change Management, 24,* 767–788.

Holts, K. (2013). Towards a taxonomy of virtual work. *Work, organization, labour and globalization, 7,* 31–50.

Howcroft, D., & Richardson, H. (2012). The back office goes global: Exploring connections and contradictions in shared service centers. *Work, Employment & Society, 26,* 111–127.

Huws, U. (2006). *The making of a cybertariat: Virtual work in a real world*. New York: Monthly Review Press.

Huws, U., Dahlmann, S., & Flecker, J. (2004). *Outsourcing of ICT and related services in the EU: A status report. Report for the European Foundation for the improvement of living and working conditions*. Luxembourg: Office for Official Publications of the European Communities.

Jackson, G., & Deeg, R. (2006). How many varieties of capitalism? Comparing the comparative institutional analyses of capitalist diversity. *MPIfG Discussion Paper 06/2*. Köln: Max-Planck-Institut für Gesellschaftsforschung.

Jürgens, K. (2006). *Arbeits- und Lebenskraft. Reproduktion als eigensinnige Grenzziehung*. Wiesbaden: VS Verlag für Sozialwissenschaften.

Kalleberg, A. L. (2011). *Good jobs—Bad jobs: The rise of polarized and precarious employment systems in the United States 1970s to 2000s*. New York: Russell Sage Foundation.

Kleemann, F., Matuschek, I., & Voß, G. G. (2002). Subjektivierung von Arbeit – ein Überblick zum Stand der soziologischen Diskussion. In M. Moldaschl & G. G. Voß (Eds.), *Subjektivierung von Arbeit* (pp. 53–100). München/Mering: Hampp.

Kraemer, K. (2010). Abstiegsängste in Wohlstandslagen. In N. Burzan & P. A. Berger (Eds.), *Dynamiken (in) der gesellschaftlichen Mitte* (pp. 201–229). Wiesbaden: VS Verlag.

Krippner, G. R. (2011). *Capitalizing on crisis. The political origins of the rise of finance*. Cambridge: Harvard University Press.

Lacity, M. C., Willcocks, L. P., & Rottman, J. W. (2008). Global outsourcing of back office services: Lessons, trends, and enduring challenges. *Strategic Outsourcing: An International Journal, 1,* 13–34.

Lohr, K. (2013). Subjektivierung von Arbeit. In H. Hirsch-Kreinsen & H. Minssen (Eds.), *Lexikon der Arbeits- und Industriesoziologie* (pp. 430–437), Berlin: Edition sigma.

Lüthje, B. (2006). Electronics contract manufacturing: Globale Produktion und neue Arbeitsregimes in China. *WSI-Mitteilungen 1/2006*.

Marchington, M., Grimshaw, D., Rubery, J., & Willmott, H. (Eds.). (2005). *Fragmenting work: Blurring organizational boundaries and disordering hierarchies*. Oxford: Oxford University Press.

Marrs, K. (2010). Herrschaft und Kontrolle in der Arbeit. In F. Böhle, G. G. Voß, & G. Wachtler (Eds.), *Handbuch Arbeitssoziologie* (pp. 331–356). Wiesbaden: VS Verlag.

Mau, S., Mewes, J., & Schöneck, N. M. (2012). What determines subjective socio-economic insecurity? Context and class in comparative perspective. *Socio-Economic Review, 10,* 655–682.

Millward, L. J., Haslam, S. A., & Postmes, T. (2007). Putting employees in their place: The impact of hot desking on organizational and team identification. *Organization Science, 18,* 547–559.

Offe, C. (1985). *Disorganized capitalism. Contemporary transformation of work and politics*. London: MIT Press.

Perlow, L. A. (1999). The time famine: Towards a sociology of work time. *Administrative Science Quarterly, 44,* 57–81.

Pfeiffer, S. (2012). Technologische Grundlagen der Entgrenzung: Chancen und Risiken. In B. Badura, A. Ducki, H. Schröder, J. Klose, & M. Meyer (Eds.), *Fehlzeiten Report 2012. Gesundheit in der flexiblen Arbeitswelt: Chancen nutzen–Risiken minimieren* (pp. 15–21). Berlin, Heidelberg: Springer.

Polanyi, M. (1966). *The tacit dimension*. Garden City, N.Y.: Doubleday.

Ramioul, M., & Van Hootegem, G. (2015). Relocation, the restructuring of the labour process and job quality. In J. Drahokoupil (Ed.), *The outsourcing challenge: Organizing workers across fragmented production networks* (pp. 91–115). Brussels: ETUI.

Reed, M. (2010). Control in contemporary work organizations. In P. Blyton, E. Heery, & P. Turnbull (Eds.), *Reassessing the employment relationship* (pp. 41–70). Basingstoke: Macmillan.

Rosa, H. (2005). *Beschleunigung. Die Veränderung der Zeitstrukturen in der Moderne*. Frankfurt: Suhrkamp.

Ross, A. (2013). In search of the lost paycheck. In: T. Scholz (Ed.), *Digital labor. The Internet as playground and factory* (pp. 13–32). New York and London: Routledge.

Sauer, D. (2010). Vermarktlichung und Vernetzung der Unternehmens- und Betriebsorganisation. In F. Böhle, G. G. Voß, & G. Wachtler (Eds.), *Handbuch Arbeitssoziologie* (pp. 545–568). Wiesbaden: VS Verlag.

Scherschel, K., Streckeisen, P., & Krenn, M. (Eds.). (2012). *Neue Prekarität Die Folgen aktivierender Arbeitsmarktpolitik - europäische Länder im Vergleich*. Frankfurt am Main: Campus Verlag.

Schmiede, R. (1996). Informatisierung, Formalisierung und kapitalistische Produktionsweise. Entstehung der Informationstechnik und Wandel der gesellschaftlichen Arbeit. In R. Schmiede (Ed.), *Virtuelle Arbeitswelten. Arbeit, Produktion und Subjekt in der „Informationsgesellschaft“* (pp. 15–47). Berlin: Edition sigma.

Schmiede, R. (2015). *Arbeit im informierten Kapitalismus. Aufsätze 1976-2015*. Berlin: Edition Sigma.

Schörpf, P., Flecker, J., Schönauer, A., & Eichmann, H. (2016). *Triangular love-hate. Governance and control in crowdworking*. Manuscript submitted to New Technology, Work and Employment.

Seifert, H. (2007). Arbeitszeit - Entwicklungen und Konflikte. *Aus Politik und Zeitgeschichte, 4–5,* 17–24.

Sørensen, O. H., Ramioul, M., & Naujanienė, R. (2015). Autonomy in unlikely places: Preconditions in low-skilled jobs. In U. Holtgrewe, V. Kirov, & M. Ramioul (Eds.), *Hard work in new jobs. The quality of work and life in European growth sectors* (pp. 231–249). Houndmills and London: Palgrave.

Sproll, M. (2016). Missing links in service value chain analysis—The case of call centers in the Brazilian banking sector. In J. Flecker (Ed.), *Space, place and global digital work*. London: Palgrave Macmillan.

Sturdy, A., Fleming, P., & Delbridge, R. (2010). Normative control and beyond in contemporary capitalism. In P. Thompson & C. Smith (Eds.), *Working life. Renewing labour process analysis* (pp. 113–135). Basingstoke: Palgrave Macmillan.

Taylor, P. (2010). The globalization of service work: Analysing the transnational call centre value chain. In: P. Thompson & C. Smith (Eds.), *Working life. Renewing labour process analysis* (pp. 244–268). Hampshire: Palgrave.

Thompson, P. (2003). Disconnected capitalism: Or why employers can't keep their side of the bargain. *Work, Employment & Society, 17,* 78–359.

Thompson, P. (2013). Financialization and the workplace: Extending and applying the disconnected capitalism thesis. *Work, Employment & Society, 27,* 472–488.

Vartiainen, M. (2006). Mobile virtual work—concepts, outcomes and challenges. In E. Andriessen & M. Vartiainen (Eds.), *Mobile virtual work. A new paradigm?* (pp. 13–44). Berlin and Heidelberg: Springer.

Voß, G. G. (1998). Die Entgrenzung von Arbeit und Arbeitskraft. Eine subjektorientierte Interpretation des Wandels der Arbeit. *Mitteilungen aus der Arbeitsmarkt- und Berufsforschung, 31,* 473–487.

Voß, G. G., & Pongratz, H. J. (1998). Der Arbeitskraftunternehmer. Eine neue Grundform der Ware Arbeitskraft? *Kölner Zeitschrift für Soziologie und Sozialpsychologie, 50,* 131–158.

Wajcman, J. (2015). *Pressed for time. The acceleration of life in digital capitalism*. Chicago: University of Chicago Press.

Watson, T. J. (2008). *Sociology, work and industry* (5th ed.). London and New York: Routledge.

Windeler, A., & Wirth, C. (2010). Netzwerke und Arbeit. In F. Böhle, G. G. Voß, & G. Wachtler (Eds.), *Handbuch Arbeitssoziologie* (pp. 539–596). Wiesbaden: VS Verlag.

Windolf, P. (Ed.). (2005). *Finanzmarkt-Kapitalismus*. Sonderheft der Kölner Zeitschrift für Soziologie und Sozialpsychologie: Analysen zum Wandel von Produktionsregimen. 45.

Windolf, P. (2008). Eigentümer ohne Risiko. Die Dienstklasse des Finanzmarkt-Kapitalismus. *Zeitschrift für Soziologie, 37,* 516–535.

Author Biographies

Jörg Flecker is a professor of sociology and chair of the sociology department at the University of Vienna, Austria. Previously, he was scientific director of the Working Life Research Institute (FORBA) in Vienna. His main research interests include changes in work, the labor market, transnationalization, digitalization, employment relations, and the political extreme right.

Theresa Fibich is a university assistant (prae doc) at the Department of Sociology at University of Vienna. Her research interests cover work and employment and family sociology. Special attention is dedicated to the transnationalization of working arrangements, the reconciliation of work and family life and the gendered division of paid and unpaid labor within households.

Klaus Kraemer is Full Professor of Applied Sociology: Economy, Organizations, Social Problems at the Department of Sociology, University of Graz. His main areas of research are economic sociology, the sociology of money and financial markets, the sociology of precarization, and social inequality.

Chapter 3
The Intensification of Work

Matea Paškvan and Bettina Kubicek

My dear, here we must run as fast as we can, just to stay in place.
And if you want to go anywhere you must run twice as fast as that!

(Carroll 1865)

Even though the quotation from Carroll's (1865) book "Alice's Adventures in Wonderland" was written more than a century ago, his words seem more than appropriate for today's world. Reframed in the words of Gleick (1999), we are now experiencing "an acceleration of just about everything", which manifests itself in the fact that people have never so little time and so many things to do. This overall acceleration (Rosa 2003) has a particularly significant effect on the working world (Korunka and Kubicek 2013). Working faster and working more seems to have become a dictum of daily activities at our jobs, making work intensification a global issue.

This chapter aims to shed more light on work intensification, its causes and consequences. In doing so, we give insights on the inherent characteristics of work intensification and the global changes triggering work intensification while providing a historical overview capturing the development of work intensification over time. After discussing possible consequences of work intensification for employees' well-being, we also present social, organizational and personal resources that can buffer the negative effects of work intensification.

M. Paškvan (✉)
Faculty of Psychology, Universitätsstraße 7, 1010 Vienna, Austria
e-mail: matea.paskvan@univie.ac.at

B. Kubicek
University of Applied Sciences Upper Austria, Hagenberg im Mühlkreis, Austria
e-mail: bettina.kubicek@univie.ac.at

C. Korunka and B. Kubicek (eds.), *Job Demands in a Changing World of Work*,
DOI 10.1007/978-3-319-54678-0_3

3.1 Characteristics of Work Intensification

Work intensification describes a process characterized by an intensification of work effort. Specifically, work intensification captures the intensive effort that is needed to complete more tasks within one working day. In this sense, it can be distinguished from extensive effort or longer work hours (Green 2001). Extensive effort does not necessarily mean that work intensifies, because the extension of work (as captured in longer work hours) does not imply that work itself intensifies; rather it could also arise from a long-hours culture (Chatzitheochari and Arber 2009). In line with this differentiation between work intensification and extensification, recent research indicates that working hours are declining (Green 2001; OECD 1999), while at the same time work intensity is increasing (Green 2001).

In sum, work intensification is defined as a multifaceted construct characterized by the need to work faster and face tighter deadlines, by a reduction of idle time and the need to conduct a number of work tasks simultaneously (Green 2004b; Kubicek et al. 2015). These core characteristics of work intensification are best illustrated by the statement of an employee who was interviewed in a large project on intensified demands (Beushausen 2014):

> That time is becoming more of an issue and now, in those weekly hours there are just more demands placed on you and then you have to make sure that you get it all done more quickly somehow. (Assistant to the executive board, 51 years old).

A recent study (Kubicek et al. 2015) builds upon these characteristics of work intensification and introduces a measure (the Intensification of job Demands Scale; i.e., IDS) enabling practitioners as well as scholars to measure intensified job demands. By applying a direct measure of change (Burchell et al. 2002) in which the item wording captures the intensification of demands (e.g., "In the last five years, it is increasingly rare to have enough time for work tasks"), the IDS provides an opportunity to capture work intensification directly via respondents' statements indicating whether they have experienced an increase in the corresponding demands. Despite the well-known advantages of longitudinal research, this research is both more efficient than longitudinal measures and is not effected by response shift (Schwartz and Sprangers 1999) or regression effects (Popper et al. 2012). Through an extensive validation process, the authors of the scale showed that their multifaceted measure of work intensification has moderate positive correlations with time pressure, showing that the IDS and especially its measure of work intensification is related to traditional job demands. However, by showing that work intensification has an effect on emotional exhaustion, depersonalization and job satisfaction beyond the traditional measure of time pressure, they ensure the innovative potential of the IDS. Focusing on the growing attention directed toward the intensification of various job demands and especially work intensification, we argue in line with the authors (Kubicek et al. 2015) that the IDS expands existing measures of work characteristics as it allows scholars to capture work intensification directly.

3.2 Global Changes and the Intensification of Work

As the first two chapters (see Korunka and Kubicek, Chap. 1 and Flecker et al., Chap. 2) impressively show, today's working world is undergoing considerable changes, with many of these changes triggering work intensification (see Fig. 3.1). Socio-economic changes such as financialization, rising (global) competition, the

Global Socio-Economic Changes

- Global trade markets
- Rising (global) competition
- Technological advancements

Change on the Organizational Level

- Reduction of hierarchy levels
- Downsizing
- Outsourcing
- Speeding up of decision-making processes and reduction of product cycles
- Implementation of flexible organizations
- New management strategies

Change on the Individual Level

- Higher performance expectations
- More control over work - from "employee" to "entreployee"
- Shorter time intervals and tighter deadlines

Work Intensification

- Increase in working speed
- More tasks within the same time period
- Reduction of idle time
- Multitasking

Fig. 3.1 Global changes and work intensification

deregulation of organizations as well as technological advancements accompanied by informatization and digitalization, to mention but a few, alter organizational contexts and in the long run challenge employers as well as employees (Landsbergis 2003; Sauter and Murphy 2003; Sparks et al. 2001).

Financialization in general and the internationalization of markets in particular lead to increasing competition not only on local markets but also internationally, which may in turn trigger an intensification of work (Olsen et al. 2010). Confronted with global competition, an organization has to react to these changes in order to remain efficient (for an overview see Flecker et al., Chap. 2). In doing so, downsizing and outsourcing are often used as managerial strategies (Cascio 2003, 2005), which leads to higher performance expectations among those employees that remain with the organization (Landsbergis 2003; Sauter and Murphy 2003; Sparks et al. 2001), triggering work intensification.

In order to remain competitive, organizations also try to speed up decision-making processes and reduce product cycles (Smith 1997), concentrate on their core competencies, and outsource other activities (see Flecker, Fibich and Kraemer, Chap. 2). Moreover, they introduce more flexible organizational structures (Cascio 2005), demanding that employees take more and more control over their own work instead of receiving direct guidance from the organization (Voß and Pongratz 1998). Notwithstanding the positive effects of more job control, increasing flexibility also implies an intensification of work (Kelliher and Anderson 2010).

In addition, the general trend to an accelerated technological change captured by ever changing and faster information and com-munication technologies does not only challenge employees' skills (see Flecker et al., Chap. 2; Korunka et al. 2014) but also increases working speed by far (Rosa 2003). One would expect that technological advancements would lead to more leisure time, however, paradoxically those advancements initiate an acceleration cycle (Rosa 2003).

Summing up, global changes do not only challenge organizations; they also alter employees' working conditions (Cascio 2003). In this vein, employees report that they have to conduct their work within shorter time intervals, as they need to deal with tighter deadlines. In order to cope with these new demands, employees feel the need to increase their working speed (e.g., Green 2004a, b), accomplish more tasks in the same or even shorter periods of time, reduce their breaks or idle times (Roberts 2007), and deal with several tasks simultaneously (multitasking; Kenyon 2008).

Hence, work intensification seems to be an integral consequence of global social-economic changes (e.g., Green 2004a; Sparks et al. 2001). Recent qualitative interviews support this view by showing that work intensification is triggered by informatization and digitalization (i.e., technological advancements; Beushausen 2014):

> …it slowly crept in and became more and more. Until of course all those electronical media came into use…just email correspondence alone that has all multiplied … (Bank employee, 54 years old)

> Yes, the transfer of information increased rapidly, of course triggered by computers, and by new media… (Bank employee, 49 years old)

> Before, he [the customer] could call me and I could just talk to that one customer because I didn't have three phones but now, well, I have more communication media by which customers or important co-workers can contact me. And this just creates much more time pressure. (Bank employee, 49 years old)

These qualitative insights, in turn, are in line with research showing that due to global changes, work intensity has increased over the last few decades (Burchell and Fagan 2004; Burchell et al. 2002; Green 2001, 2004b; Parent-Thirion et al. 2012; Sauter and Murphy 2003; Sparks et al. 2001; Valeyre 2004).

3.3 The Intensification of Work – Now and Then

The intensification of work and acceleration in general are not new at all and have always been a part of history; however, the rate of social change has accelerated (Rosa 2003). To shed more light on work intensification and its development over time, the following section presents a historical overview of work intensification beginning in the 1990s.

An intensification of work was already observable in the 1990s. Analyzing a trend dataset, namely the European Working Conditions Survey, Paoli and Merllié (2005) reported an intensification of work from 1990 to 2000, with a peak between 1990 and 1995. Relying on the same dataset, Green and McIntosh (2001) focused additionally on national differences in work intensification. By doing so, they report the strongest increase in work intensity in Great Britain followed by Ireland and France, while the lowest rates of intensification were reported in West Germany. Analyzing another international panel dataset (International Social Survey Programme), Olsen et al. (2010) show an increase in work intensity in West Germany, Norway, Great Britain and the United States in at least one period of time (1989, 1997 and 2005). All of these results are based on datasets that do not directly measure work intensification; rather they ask employees about their work intensity and compare employees' self-reported levels of work intensity at different points in time. Other pieces of research directly asking employees whether they have experienced work intensification, however, report similar results. Burchell and colleagues (2002) found that over 60% of respondents reported an intensification of work in Great Britain. In sum, researchers have identified an intensification of work in Europe and the United States in the 1990s, with the highest levels in Great Britain. It is hypothesized that those increases in work intensity go along with socio-economic changes such as technological advancements and the globalization of trade markets (Green 2004a, b; Green and McIntosh 2001).

A recent study builds upon this research, investigating whether the intensification of work remained in place up to the first decade of the 21st century. Again analyzing the European Working Conditions Survey, the authors show that an increase in work intensity occurred in Germany, Spain, and France from 2000 until 2010 (Kubicek et al. 2014). In contrast, there was no general trend toward an intensification of work observable in Finland, and a decline rather than an increase in work intensity was visible in the United Kingdom. In order to model individual trajectories, the authors additionally analyzed panel datasets of work intensity over time. They report that different patterns of work intensity groups are observable in Germany (based on data from the German Socio-Economic Panel) and in the United States (based on data from the Health and Retirement Study). In Germany as well as in the United States, most employees belong to rather stable groups. Whereas in Germany 50% report a rather stable but high work intensity pattern over time, in the United States 53% report rather stable but low levels of work intensity. These differences may be due to the fact that the German dataset is representative, whereas the US dataset is based on a cohort of relatively older people (Health and Retirement Study). Aside from those stable groups, results from both datasets indicate that there are specific groups of employees who have indeed experienced an increase in work intensity. In Germany, around 33% of employees reported an increase, whereas in the United States 36% of employees indicated that they had experienced an increase in work intensity over time. Interestingly the trajectories of work intensity were linked to trajectories of computer usage: increases in work intensity were accompanied by increases in computer usage. Again these results indicate that work intensification is linked to technological advancements. Summing up, recent findings indicate that the trend toward work intensification is far from being universal; however, results support the view that work intensification is a central issue for employees in modern Western societies.

3.4 The Effects of Work Intensification on Well-Being and Work Outcomes

Work and organizational psychologists are especially interested in the effects of work characteristics on employees. They propose different labels for characterizing working conditions and different models for capturing their positive and negative impacts on employees. Three of the most important models will be used to elaborate on work intensification and its effects on employees. First, the Job Demands-Resources model (JD-R model, Bakker and Demerouti 2007) will be presented and substantiated with the compensatory regulation control model (Hockey 1997) to explain the effects of work intensification on employee strain and well-being. Second, the challenge-hindrance framework (e.g., Cavanaugh et al. 2000) will be used to explain its effects on employee well-being as well as on work-related outcomes such as job satisfaction. Third, Lazarus' cognitive appraisal theory will be used to further explain why work intensification impairs well-being and job satisfaction.

3.4.1 Work Intensification Within the Job Demands-Resources Model

According to the JD-R model (Bakker and Demerouti 2007), a distinction needs to be made between two general categories of work characteristics that have different effects on employees' well-being and influence work outcomes such as job satisfaction and performance in different ways.

The first category of work characteristics is labeled job demands. "Job demands refer to those physical, psychological, social, or organizational aspects of the job that require sustained physical and/or psychological (cognitive and emotional) effort or skills and are therefore associated with certain physiological and/or psychological costs" (Bakker and Demerouti 2007: 312). Such demands include time pressure, emotional demands, contradictory demands, physical demands and work effort in general (e.g., Bakker and Demerouti 2007; Karasek 1979; Siegrist 1996). Job demands describe characteristics of the job and might be different for every occupation (Bakker and Demerouti 2007).

In contrast to job demands, the second category, namely resources, describes positive aspects of the working environment. Job resources have several positive effects on employees: they motivate employees to achieve work goals, and might also reduce the negative effects of job demands (this aspect will be described in further detail in the following section). Moreover, in line with self-determination theory (Deci and Ryan 2000), resources stimulate personal growth, learning and personal development (Bakker and Demerouti 2007). Such resources may stem from the person (e.g., self-esteem, initiative, etc.), from the social environment (e.g., supervisor support, climate, etc.), as well as from the organization itself (e.g., participative decision-making, etc.).

According to the JD-R model (Bakker and Demerouti 2007), work intensification ought to be conceptualized as a demand as it requires sustainable effort and is associated with certain physiological and psychological costs on the part of employees (Green 2004b). In contrast to the negative consequences work intensification holds for employees, organizations seem to benefit from work intensification. Side effects of work intensification such as increased work effort boost the productivity of an organization and ensure the economic profitability of companies (e.g., Green 2001; Valeyre 2004). However, the effects of work intensification on employees are mostly negative, which calls the long-term benefits of work intensification for organizations into question. Prior studies have shown that work intensification is associated with strain indicators (Allan et al. 1999; Green 2001), reduced work-home balance (Macky and Boxal 2008), heightened work-home conflict (Kubicek and Tement 2016) and a reduction in employees' job satisfaction (Brown 2012; Green 2004b). Those results are given further weight by a recent longitudinal study reporting that increases in work intensification lead to higher levels of emotional exhaustion and lower levels of engagement (Korunka et al. 2014). In sum, the positive effects of work intensification for organizations are rather short lived. In the long run, organizations suffer from the consequences of work intensification due to reduced employee engagement and increased costs due to absenteeism.

Having shown that work intensification harms employees' well-being, the question arises as to how such negative effects emerge. The compensatory regulation control model (Hockey 1997) describes two different levels of regulation that determine human performance in the face of high demands and may provide an explanation for the underlying psychological mechanisms that link work intensification to strain and reduced employee well-being.

Active coping means that employees increase their effort in order to reach increased performance goals. Increased effort in turn is associated with physical (increased heart rate, increased adrenalin, and higher blood pressure) and psychological (fatigue) costs. Employees facing only short term increases in demands try to regulate their behavior via increased work effort and/or neurophysiological activation in order to meet those demands. However, if they consistently face higher demands, as is the case with work intensification, increasing work effort and activation will lead to a depletion of their resources. Employees facing work intensification need to perform more tasks during the workday and also reduce their idle times, which in turn might lead to work overload (Burchell et al. 2002) and hamper recovery on the job. Thus, high effort accompanied with insufficient recovery, as is the case for employees facing work intensification, may lead to emotional and/or physical exhaustion in the long run (Meijman and Mulder 1998).

Passive coping is another regulation mode for meeting high demands (Hockey 1997). Passive coping means that employees adjust their goals rather than their work effort. A reduction in work accuracy, for instance, may help to meet higher work expectations. In contrast to active coping, passive coping is not accompanied by physical and psychological costs; however, it also has downsides. Complete disengagement from work goals and work itself as well as a reduction in work quality and more processing errors might be the consequences of passive coping. Qualitative research findings support these assumptions (Beushausen 2014). An assistant, for instance, reported that she reacted to increased work demands in the following way: "… you have to make sure to work it all off much more quickly somehow. I, personally, manage to do this but it is just much more prone to errors" (Assistant to the executive board, 51 years old). The following account shows complete disengagement from work. A bank employee admitted that in order to reach the performance goals set by upper management, he and his supervisor and colleagues indicated having more appointments than they actually had with costumers:

> Partially, I have to say, I can say it now anyway, we faked appointments with our supervisors. It's nearly impossible to schedule six appointments a day and besides you have your daily work and you have to provide good service to costumers, that is just not possible. (Bank employee, 28 years old).

In addition to these statements illustrating passive coping when facing work intensification, empirical studies also find support for passive coping mechanisms. Hence, research indicates that work intensification and its associated antecedents (downsizing, etc.) are accompanied by increases in occupational accidents and injuries (e.g., Landsbergis 2003). Thus, even if the passive coping mechanisms are not directly associated with physical and psychological costs, negative effects occur in the long run for employees and organizations.

Summing up, in line with the JD-R model (Bakker and Demerouti 2007), work intensification ought to be conceptualized as a demand that has mostly negative effects on employees. Employees use different coping mechanisms when faced with work intensification; however, both mechanisms have their downsides: strain and reduced well-being on the one hand, and reduced productivity, errors, accidents or work disengagement on the other hand.

Recent theoretical developments point to a need to differentiate between various demands, as they have various effects for employees. In order to put work intensification in line with these recent theoretical developments in work and organizational psychology, the following section introduces the challenge-hindrance framework and applies this framework to work intensification.

3.4.2 Work Intensification Within the Challenge-Hindrance Framework

The JD-R model (Bakker and Demerouti 2007) was recently extended because empirical and theoretical evidence indicates that not all demands are equal, and thus a solely negative view of demands needs to be called into question (Cavanaugh et al. 2000). The challenge-hindrance framework (e.g., Cavanaugh et al. 2000) builds upon these ideas distinguishing between two types of demands, namely challenge and hindrance demands (Crawford et al. 2010; Van den Broeck et al. 2010).

Hindrance demands are seen as thwarting personal growth and goal attainment and promoting strain (Cavanaugh et al. 2000; Podsakoff et al. 2007). In line with this assumption, a meta-analysis (Podsakoff et al. 2007) revealed that hindrance demands are associated with increased strain and decreased job satisfaction. Challenge demands, in contrast, are also considered to promote strain. But in addition to this energy-depleting effect, they are also assumed to be motivating and to foster personal growth and goal attainment (Cavanaugh et al. 2000; LePine et al. 2005). Thus, whereas hindrance demands only entail negative effects for employees, challenge demands promote motivation and personal growth (for instance by increasing self-esteem) but simultaneously drain employees' energy (Widmer et al. 2012).

In transferring these considerations to work intensification, work intensification should be classified as a hindrance demand, as it is associated with increased strain

(Allan et al. 1999; Green 2001) and with reduced motivational states such as decreased job satisfaction (Brown 2012; Green 2004b) and engagement (Korunka et al. 2014). However, Aubert (2009) reports that managers facing work intensification not only experience strain but also mentions that work intensification brings about a thrill accompanied by a sense of delight. Another interview study also supports the view that the effects of work intensification are ambivalent (Beushausen 2014). A 49-year-old bank employee, for instance, says:

> [talking about fast paced information processes] …yes. Whether it is an email, a to do-list which we have, order portals and all. This is crazy! Regarding this we had the most changes, Yes. I mean, the flow of information is so fast paced and of course you have to translate everything into action rapidly too, which is positive on one hand but naturally it is also a major strain.

The positive effects associated with work intensification seem however to be rather short-lived. In the long run, the managers under study experienced less control over their work situation, leading to irritation or depressive symptoms (Aubert 2009). However, it may still be the case that work intensification can sometimes be seen by employees as a challenge. In order to resolve the question of whether work intensification should be conceptualized as a challenge or hindrance, a recent study measured work intensification via direct appraisal (Paškvan et al. 2016).

3.4.3 The Cognitive Appraisal of Work Intensification

Although the challenge-hindrance framework (Cavanaugh et al. 2000) builds upon cognitive appraisal theory (e.g., Lazarus and Folkman 1984), the tradition of measuring cognitive appraisal directly was long considered to be a bit outdated – although it has received a lot of attention lately (Webster et al. 2011). This move towards directly measuring appraisal in order to directly classify demands as challenge or hindrance demands is important as recent studies show that based on the challenge-hindrance framework, time pressure is defined as a classic and the most prominent challenge demand (Podsakoff et al. 2007; Widmer et al. 2012). In line with these studies empirical results solely measuring challenge appraisal, show that time pressure is appraised as a challenge (Ohly and Fritz 2010). In contrast, another study directly measuring eldercare workers' appraisals of time pressure in terms of hindrance and challenge appraisal (Bakker and Sanz-Vergel 2013) shows that time pressure tends to be appraised as a hindrance. Given these differences in research results, it seems advisable to measure appraisals directly before categorizing demands as challenges or hindrances (Webster et al. 2011).

Following this recommendation, the authors directly measured appraisals of work intensification (Paškvan et al. 2016), showing that in line with prior findings, work intensification is indeed appraised as a hindrance. This appraisal did not only classify work intensification as a hindrance; but the authors also showed that in addition to the energetic processes linking work intensification to reduced well-being described previously, cognitive appraisal itself is an underlying explanatory process (e.g., Lazarus and Folkman 1984). More concretely, the cognitive appraisal of work intensification explained the effect of work intensification on emotional exhaustion as well as job satisfaction (Paškvan et al. 2016).

In brief, it has been shown that work intensification has mostly negative effects on employees' well-being and work-related outcomes. Those results are underscored by the fact that work intensification is also appraised as a hindrance by employees. The cognitive appraisal process (Lazarus and Folkman 1984) as well as coping mechanisms in dealing with demands (i.e., active vs. passive coping; Hockey 1997) can explain how work intensification is linked to its outcomes.

3.5 Strategies and Resources in Dealing with Work Intensification

As mentioned above, resources are important to mitigate the negative effects of demands. Resources may stem from the person, the social environment or from the organization (Bakker and Demerouti 2007). Which resources help employees deal with the negative effects of demands depends on the inherent characteristics of the demands themselves. Resources in general may not only influence the effects of work intensification on its outcomes, they may also have an effect on the appraisal of work intensification. Indeed, the cognitive appraisal process is described as a dynamic process which may be influenced by person- and situation-specific factors (Lazarus and Folkman 1984).

In the following sections, we discuss several resources that have the potential to influence the appraisal of work intensification and reduce the negative effects of work intensification on its outcomes.

3.5.1 What Can Organizations Do? Job Resources and Organizational Climate

Within work and organizational psychology, job control (i.e., autonomy) is referred to as one of the most important resources (e.g., Karasek 1979). Employees who have job control are given discretion in deciding when, how and with what means they pursue their work tasks. Even if the solely positive view of job control has been called into question nowadays (for more information, see Kubicek et al., Chap. 4), job control may aid employees in dealing with work intensification. A high degree of control gives employees the feeling that they can directly influence their work environment (Spector 1986). As a result, control has a considerable impact on the cognitive appraisal process (Peeters et al. 1995; Terry 1991). Thus, in order to influence the cognitive appraisal process, organizations can enhance employees' control over their work situation. Moreover, a meta-analytical study on the benefits of interventions shows that individual-level (cognitive-behavioral) interventions had the greatest effects when employees had job control (Van der Klink et al. 2001). Summing up, it seems that job control is an important resource for employees faced with work intensification. This is in line with a recent study arguing that professionals need more control over the time and timing of work in order to reduce time strains (Moen et al. 2013).

According to Paškvan et al. (2016) event uncertainty and a perceived lack of control (as previously discussed) are the main drivers of an employee appraising a demand as threatening (Bordia et al. 2004; Lazarus and Folkman 1984). Instead of solely increasing the job control of each employee directly, organizations could also provide a climate in which uncertainty is reduced and personal control is enhanced. Such a climate is best described as a participative climate. Organizations offering a participative climate share information (Sagie and Koslowsky 1996; Van de Heuvel et al. 2013) and consider employees' opinions in decision-making (Tesluk et al. 1999). In fact, it has been shown that a participative climate buffers the negative effects of work intensification on its hindrance appraisal (Paškvan et al. 2016). Thus, organizations providing a participative climate could help their employees reduce the negative effects of work intensification, as employees have sufficient information to know what might happen and to estimate the likelihood of an event (Lazarus and Folkman 1984), but also the opportunity to share influence in decision-making (Mitchell 1973).

In addition to offering job control or a participative climate, organizations can help employees deal with work intensification by promoting a culture that separates work and home life. Literature on boundary management indicates that the work and home domains are delineated by physical, temporal, and psychological/mental borders (Ashforth et al. 2000; Clark 2000) and that organizations can either promote a segmentation or an integration of the work and home domains (Kreiner 2006). Work-home segmentation is characterized by a clear separation between the work and home domains, a rigid and inflexible construction of boundaries between work and home roles and rare cross-role interruptions. Drawing on the JD-R model (Bakker and Demerouti 2007), an organization's culture of work-home segmentation may be considered a job resource that protects employees from the costs of work intensification (e.g., Kubicek and Tement 2016). According to the JD-R model, job resources provide protection against demanding work situations by buffering the negative effects of job demands, especially when job demands are high (Bakker et al. 2007). Amstad and Semmer (2009), for instance, argue that work-home segmentation is important for preventing negative outcomes arising from high job demands. In line with this reasoning, Kubicek and Tement (2016) found that employees facing work intensification reported greater gains from work with regard to positive mood when their organizations promoted a separation between the work and home domains and thus allowed their employees to detach from job demands while at home.

3.5.2 WE Will Manage It! Social Support and Work Intensification

Social resources such as emotional and instrumental support from colleagues and supervisors should also help employees cope with work intensification. A climate of trust, advice, and encouragement is an important resource to help employees deal with highly demanding jobs (Hobfoll 2002). In addition to these rather passive aspects of support, active assistance such as taking over an additional task or providing direct support during periods of intense workload may be helpful for employees. Indeed, employees who receive such supports perceive fewer negative consequences from a demanding job compared to employees who do not receive social supports (Griffin et al. 2007; Karasek and Theorell 1990).

This buffering effect of social resources was analyzed in a study capturing the consequences of social acceleration and work intensification in elder care (Kubicek et al. 2013). The results confirmed that eldercare workers who experience high work intensification but at the same time work in an organization with a good social climate are more satisfied with their job than eldercare workers working in an organization with a negative social climate. These results are especially interesting given the fact that the social climate was rated by supervisors. In sum, it seems that direct help as well as a good social climate are important resources for mitigating the negative effects of work intensification on employee well-being and job satisfaction.

3.5.3 Is It All About Managing One's Self? Self and Time Management

Scholars' as well as practitioners' interest in self-management has increased over the last few years. This is due to the fact that self-management is seen as key to staying competitive, particularly in times of change (Walton 1985). Self-managing implies that behaviors previously ascribed to management become employees' responsibility, such as setting goals, self-monitoring and self-evaluation of performance (goals), self-setting corrective actions if necessary, self-penalties and self-reinforcement (Manz and Sims 1984). All of these behaviors could potentially serve as a resource for employees when facing work intensification. Employees who neatly plan out their goals and monitor their actions, for instance, could benefit from these behaviors because they perceive greater feelings of control while working under tight deadlines. Being aware of one's own goals and having a sense of work overload are key personal characteristics that help employees deal with rising demands such as work intensification and thus should be fostered by organizations (Ducki 2002).

Focusing particularly on the temporal aspect of work intensification, time management could serve as key resource. Moen et al. (2013) analyzed how higher status professionals are responding to increasingly common and chronic stress. The authors show that four time strategies are particularly common for organizing time strains: prioritizing time, scaling back obligations, blocking out time in order to engage in non-work or work issues, and time shifting are considered specific time strategies that could help employees organize their work as well as family issues (Moen et al. 2013). Thinking on the negative effects of work intensification on work-family conflict (Kubicek and Tement 2016), these strategies could also serve as resources in conditions of high work intensification. Allowing employees to shift the times and places they work to create a better fit between work and personal life or taking time off after intensive work phases could reduce the negative effects of work intensification. Studies analyzing the effects of time management, however, show mixed results. Some indicated that time management training has a limited effect on actual time management behavior (Macan 1994), while others found an effect of time management training on actual self-reported behavior (Häfner and Stock 2010), on perceived control over time (Macan 1996) and on well-being and job satisfaction (Häfner and Stock 2010; Macan 1994; Van Eerde 2003). Summing up, even if time management is hard to teach, it has a positive impact on employees. Thus, prioritizing, structuring the work day and monitoring should be seen as helpful strategies for employees facing work intensification.

Despite the importance of individual interventions, the sole focus should not be on personal interventions alone (Ducki 2002). Meta-analytical studies show that interventions with a focus on personal behavior (e.g., self-management) tend to have rather small effects (Bamberg and Busch 2006; Van der Klink et al. 2001). Thus, we argue that interventions to help employees deal with work intensification should include both organizational and personal measures and regulations. We argue in line with others (Van der Klink et al. 2001) that an individually-tailored intervention should be combined with organizational ones.

3.6 Outlook

This chapter argues that global socio-economic changes such as technological advancement, outsourcing and downsizing are drivers of work intensification. In many organizations, downsizing has been seen as a quick move to reduce costs, but the associated improvements are probably only temporary. In the long run, employees have more work to do, causing work intensification (Kubicek et al. 2015; Sauter and Murphy 2003), stress-related illness and occupational accidents (Landsbergis 2003). Thus, even if work intensification is associated with increased productivity, over time it seems that employees reporting work intensification experience lower well-being, less job satisfaction and more work-family conflict. Thus, we encourage organizations to weigh the short-term benefits of work intensification against the long-term costs associated with work intensification and its consequences. From a societal and practical perspective, downsizing as a managerial strategy should be regarded with suspicion.

This chapter also highlights the importance of job resources in cases where workers face work intensification. Work intensification as a consequence of socio-economic changes affects many employees across a wide range of organizations: hence, employers should provide sufficient resources and appropriate trainings in order to reduce the negative effects of work intensification.

References

Allan, C., O'Donell, M., & Peetz, D. (1999). Three dimensions of labor utilization: job broadening, employment insecurity and work intensification. *Current Research in Industrial Relations, 1,* 13–24.

Amstad, F. T., & Semmer, N. K. (2009). Recovery and the work-family interface. In P. Perrewé, D. C. Ganster, & S. Sonnentag (Eds.), *Research in occupational stress and well-being: Current perspectives on job-stress recovery* (pp. 125–166). Bingley, UK: Emerald.

Ashforth, B. E., Kreiner, G. E., & Fugate, M. (2000). All in a day's work: Boundaries and micro role transitions. *Academy of Management Review, 25,* 472–491.

Aubert, N. (2009). Dringlichkeit und Selbstverlust in der Hypermoderne. In: V. King & B. Gerisch (Hrsg.). *Zeitgewinn und Selbstverlust. Folgen und Grenzen der Beschleunigung.* Frankfurt/Main: Campus.

Bakker, A. B., & Demerouti, E. (2007). The Job Demands-Resources model: State of the art. *Journal of Managerial Psychology, 22*(3), 309–328. doi:10.1108/02683940710733115

Bakker, A. B., & Sanz-Vergel, A. I. (2013). Weekly work engagement and flourishing: The role of hindrance and challenge job demands. *Journal of Vocational Behavior, 83,* 397–409. doi:10.1016/j.jvb.2013.06.008

Bakker, A. B., Hakanen, J. J., Demerouti, E., & Xanthopoulou, D. (2007). Job resources boost work engagement, particularly when job demands are high. *Journal of Educational Psychology, 99,* 274–284.

Bamberg, E., & Busch, C. (2006). Stressbezogene Interventionen in der Arbeitswelt. *Zeitschrift für Arbeits-und Organisationspsychologie A&O, 50*(4), 215–226.

Beushausen, N. K. (2014). *Selbstkontrolle und Arbeitsintensivierung als neue, beschleunigende Anforderungen in berufsbiographischen Interviews* [Self-control and work intensification as new, accelerating demands in career biography interviews] (Master's thesis, University of Vienna, Vienna, Austria). Retrieved from http://othes.univie.ac.at/32479/

Bordia, P., Hobman, E., Jones, E., Gallois, C., & Callan, V. J. (2004). Uncertainty during organizational change: Types, consequences and management strategies. *Journal of Business and Psychology, 18,* 507–531. doi:10.1023/B:JOBU.0000028449.99127.f7

Brown, M. (2012). Responses to work intensification: Does generation matter? *The International Journal of Human Resource Management, 23*(17), 3578–3595.

Burchell, B., & Fagan, C. (2004). Gender and the intensification of work: Evidence from the "European Working Conditions Surveys". *Eastern Economic Journal, 30*(4), 627–642.

Burchell, B., Ladipo, D., & Wilkinson, F. (Eds.). (2002). *Job insecurity and work intensification.* New York: Routledge.

Carroll, L (Charles Lutwidge Dodgson). (1865). *Alice's Adventures in Wonderland.*

Cascio, W. F. (2003). Changes in workers, work, and organizations. In W. C. Borman, D. R. Ilgen, & R. J. Klimoski (Eds.), *Industrial and organizational psychology* (Vol. 12, pp. 401–422). New York: Wiley.

Cascio, W. F. (2005). Strategies for responsible restructuring. *The Academy of Management Executive, 19*(4), 39–50.

Cavanaugh, M. A., Boswell, W. R., Roehling, M. V., & Boudreau, J. W. (2000). An empirical examination of self-reported work stress among U.S. managers. *Journal of Applied Psychology, 85*(1), 65–74. doi:10.1037//0021-9010.85.1.65

Chatzitheochari, S., & Arber, S. (2009). Lack of sleep, work and the long hours culture: Evidence from the UK Time Use Survey. *Work, Employment & Society, 23*(1), 30–48. doi:10.1177/0950017008099776

Clark, S. C. (2000). Work/family border theory: A new theory of work/family balance. *Human Relations, 53,* 747–770.

Crawford, E. R., LePine, J. A., & Rich, B. L. (2010). Linking job demands and resources to employee engagement and burnout: A theoretical extension and meta-analytic test. *Journal of Applied Psychology, 95,* 834–848.

Deci, E. L., & Ryan, R. M. (2000). The "What" and "Why" of goal pursuits: Human needs and the self-determination of behavior. *Psychological Inquiry, 11,* 227–268.

Ducki, A. (2002). Betriebliche Gesundheitsförderung und Neue Arbeitsformen—Aktuelle Tendenzen in Forschung und Praxis. *Gruppendynamik und Organisationsberatung, 33*(4), 419–436.

Gleick, J. (1999). *Faster. The acceleration of just about everything*. Pantheon, New York.

Green, F. (2001). It's been a hard day's night: The concentration and intensification of work in late twentieth-century Britain. *British Journal of Industrial Relations, 39*(1), 53–80. doi:10.1111/1467-8543.00189

Green, F. (2004a). Why has work effort become more intense? *Industrial Relations, 43*(4), 709–741. doi:10.1111/j.0019-8676.2004.00359.x

Green, F. (2004b). Work intensification, discretion, and the decline in well-being at work. *Eastern Economic Journal, 30*(4), 615–625.

Green, F., & McIntosh, S. (2001). The intensification of work in Europe. *Labour Economics, 8*(2), 291–308. doi:10.1016/S0927-5371(01)00027-6

Griffin, J. M., Greiner, B. A., Stansfeld, S. A., & Marmot, M. (2007). The effect of self-reported and observed job conditions on depression and anxiety symptoms: A comparison of theoretical models. *Journal of Occupational Health Psychology, 12,* 334–349.

Häfner, A., & Stock, A. (2010). Time management training and perceived control of time at work. *The Journal of Psychology, 144*(5), 429–447.

Hobfoll, S. E. (2002). Social and psychological resources and adaptation. *Review of General Psychology, 6,* 307–324.

Hockey, G. R. J. (1997). Compensatory control in the regulation of human performance under stress and high workload: A cognitive-energetical framework. *Biological Psychology, 45,* 73–93.

Karasek, R., & Theorell, T. (1990). *Healthy work: Stress, productivity and the reconstruction of working life*. New York: Basic Books.

Karasek, R. A., Jr. (1979). Job demands, job decision latitude, and mental strain: Implications for job redesign. *Administrative Science Quarterly, 24,* 285–308.

Kelliher, C., & Anderson, D. (2010). Doing more with less? Flexible working practices and the intensification of work. *Human Relations, 63,* 83–106. doi:10.1177/0018726709349199

Kenyon, S. (2008). Internet use and time use: The importance of multitasking. *Time & Society, 17* (2–3), 283–318.

Korunka, C., & Kubicek, B. (2013). Beschleunigung im Arbeitsleben: Neue Anforderungen und deren Folgen [Acceleration in working life. New demands and their consequences]. In M. Morschhäuser & G. Junghanns (Eds.), *Immer schneller, immer mehr - Psychische Belastungen bei Wissens- und Dienstleistungsarbeit* (pp. 17–39). Wiesbaden: Springer.

Korunka, C., Kubicek, B., Paškvan, M., & Ulferts, H. (2014). Changes in work intensification and intensified learning: Challenge or Hindrance demands? *Journal of Managerial Psychology, 30* (7), 786–800.

Kreiner, G. E. (2006). Consequences of work-home segmentation or integration: A person-environment fit perspective. *Journal of Organizational Behavior, 27,* 485–507.

Kubicek, B., & Tement, S. (2016). Work intensification and the work-home interface: The moderating effect of individual work-home segmentation strategies and organizational segmentation supplies. *Journal of Personnel Psychology, 15,* 76–89. doi:10.1027/1866-5888/a000158

Kubicek, B., Korunka, C., Paškvan, M., Prem, R., & Gerdenitsch, C. (2014). Changing working conditions at the onset of the twenty-first century: Facts from international datasets. In C. Korunka & P. Hoonakker (Eds.), *The impact of ICT on quality of working life* (pp. 25–41). Netherlands: Springer.

Kubicek, B., Paškvan, M., & Korunka, C. (2013, Mai). *How to stay satisfied when work intensity increases? Team social climate helps.* Poster presented at Work, Stress and Health Congress, Los Angeles, USA.

Kubicek, B., Paškvan, M., & Korunka, C. (2015). Development and validation of an instrument for assessing job demands arising from accelerated change: The intensification of job demands scale (IDS). *European Journal of Work and Organizational Psychology.* doi:10.1080/1359432X.2014.979160

Landsbergis, P. A. (2003). The changing organization of work and the safety and health of working people: A commentary. *Journal of Occupational and Environmental Medicine, 45*(1), 61–72.

Lazarus, R. S., & Folkman, S. (1984). *Stress, appraisal and coping*. New York: Springer.

LePine, J. A., Podsakoff, N. P., & LePine, M. A. (2005). A meta-analytic test of the challenge stressor–hindrance stressor framework: An explanation for inconsistent relationships among stressors and performance. *The Academy of Management Journal, 48*(5), 764–775. doi:10.5465/AMJ.2005.18803921

Macan, T. H. (1994). Time management: Test of a process model. *Journal of Applied Psychology, 79*(3), 381.

Macan, T. H. (1996). Time-management training: Effects on time behaviors, attitudes, and job performance. *The Journal of Psychology, 130*(3), 229–236.

Macky, K., & Boxall, P. (2008). High-involvement work processes, work intensification and employee well-being: A study of New Zealand worker experiences. *Asia Pacific Journal of Human Resources, 46*(1), 38–55.

Manz, C. C., & Sims, H. P. (1984). Searching for the "unleader": Organizational member views on leading self-managed groups. *Human Relations, 37*(5), 409–424.

Meijman, T. F., & Mulder, G. (1998). Psychological aspects of workload In P. J. D. Drenth, H. Thierry, & C. J. de Wolff (Eds.), *Handbook of work and organizational psychology: Work psychology* (2nd ed., pp. 5–33). Hove, U.K.: Psychology Press.

Mitchell, T. R. (1973). Motivation and participation: An integration. *Academy of Management Journal, 16*(4), 670–679. doi:10.2307/254699

Moen, P., Lam, J., Ammons, S., & Kelly, E. L. (2013). Time work by overworked professionals strategies in response to the stress of higher status. *Work and Occupations, 40*(2), 79–114.

OECD. (1999). *Implementing the OECD job strategy: Assessing performance and policy.* Paris: OECD.

Ohly, S., & Fritz, C. (2010). Work characteristics, challenge appraisal, creativity and proactive behavior: A multi-level study. *Journal of Organizational Behavior, 31,* 543–565. doi:10.1002/job.633

Olsen, K. M., Kalleberg, A. L., & Nesheim, T. (2010). Perceived job quality in the United States, Great Britain, Norway and West Germany, 1989-2005. *European Journal of Industrial Relations, 16,* 221–240. doi:10.1177/0959680110375133

Paoli, P., & Merllié, D. (2005). *Eurofound.* Luxembourg: Ten years of working conditions in the European Union.

Parent-Thirion, A., Vermeylen, G., van Houten, G., Lyly-Yrjänäinen, M., B., I., Cabrita, J., & Niedhammer, I. (2012). *Fifth European working conditions survey.* Luxembourg: Publications Office of the European Union.

Paškvan, M., Kubicek, B., Prem, R., & Korunka, C. (2016). Cognitive appraisal of work intensification. *International Journal of Stress Management, 23*(2), 124–146.

Peeters, M. C. W., Buunk, B. P., & Schaufeli, W. B. (1995). A micro-analytic exploration of the cognitive appraisal of daily stressful events at work: The role of controllability. *Anxiety, Stress, and Coping, 8*(2), 127–139. doi:10.1080/10615809508249369

Podsakoff, N. P., LePine, J. A., & LePine, M. A. (2007). Differential challenge stressor-hindrance stressor relationships with job attitudes, turnover intentions, turnover, and withdrawal behavior: A meta-analysis. *Journal of Applied Psychology, 92*(2), 438–454. doi:10.1037// 0024-9010.92.2.438

Popper, V., Spiel, C., & von Eye, A. (2012). Evaluation von Führungskräfteentwicklung: Lösungsansätze zur Sicherung methodischer Standards an einem Fallbeilspiel [Evaluation of leadership development: Approaches for assuring methodological standards based on a case study]. *Zeitschrift für Evaluation, 11,* 39–59.

Roberts, K. (2007). Work-life balance—The sources of the contemporary problem and the probable outcomes. *Employee Relations, 29*(4), 334–351.

Rosa, H. (2003). Social acceleration: Ethical and political consequences of a desynchronized high-speed society. *Constellations, 10,* 3–33.

Sagie, A., & Koslowsky, M. (1996). Decision type, organisational control, and acceptance of change: An integrative approach to participative decision making. *Applied Psychology, 45*(1), 85–92.

Sauter, S. L., & Murphy, L. R. (2003). Monitoring the changing organization of work: International practices and new developments in the United States. *Sozial-und Präventivmedizin, 48*(6), 341–348. doi:10.1007/s00038-003-3026-2

Schwartz, C. E., & Sprangers, M. A. G. (1999). Methodological approaches for assessing response shift in longitudinal health-related quality-of-life research. *Social Science and Medicine, 48,* 1531–1548.

Siegrist, J. (1996). Adverse health effects of high-effort/low-reward conditions. *Journal of Occupational Health Psychology, 1*(1), 27.

Smith, V. (1997). New forms of work organization. *Annual Review of Sociology, 23,* 315–339.

Sparks, K., Faragher, B., & Cooper, C. L. (2001). Well-being and occupational health in the 21st century workplace. *Journal of Occupational and Organizational Psychology, 74*(4), 489–509. doi:10.1348/096317901167497

Spector, P. E. (1986). Perceived control by employees: A meta-analysis of studies concerning autonomy and participation at work. *Human Relations, 39,* 1005–1016.

Terry, D. J. (1991). Coping resources and situational appraisals as predictors of coping behavior. *Personality and Individual Differences, 12*(10), 1031–1047. doi:10.1016/0191-8869(91) 90033-8

Tesluk, P. E., Vance, R. J., & Mathieu, J. E. (1999). Examining employee involvement in the context of participative work environments. *Group and Organization Management, 24*(3), 271–299. doi:10.1177/1059601199243003

Valeyre, A. (2004). Forms of work intensification and economic performance in French manufacturing. *Eastern Economic Journal, 30*(4), 643–658.

Van de Heuvel, M., Demerouti, E., Bakker, A. B., & Schaufeli, W. B. (2013). Adapting to change: The value of change information and meaning-making. *Journal of Vocational Behavior, 83*(1), 11–21. doi:10.1016/j.jvb.2013.02.004

Van den Broeck, A., De Cuyper, N., & De Witte, H. (2010). Not all job demands are equal: Differentiating job hindrances and job challenges in the job demands-resources model. *European Journal of Work and Organizational Psychology, 19*(6), 735–759. doi:10.1080/ 13594320903223839

Van der Klink, J. J., Blonk, R. W., Schene, A. H., & Van Dijk, F. J. (2001). The benefits of interventions for work-related stress. *American Journal of Public Health, 91*(2), 270.

Van Eerde, W. V. (2003). Procrastination at work and time management training. *The Journal of psychology, 137*(5), 421–434.

Voß, G. G., & Pongratz, H. J. (1998). Der Arbeitskraftunternehmer. *Kölner Zeitschrift für Soziologie und Sozialpsychologie, 50*(1), 131–158.

Walton, R. E. (1985). From control to commitment in the workplace.

Webster, J. R., Beehr, T. A., & Love, K. (2011). Extending the challenge-hindrance model of occupational stress: The role of appraisal. *Journal of Vocational Behavior, 79*(2), 505–516. doi:10.1016/j.jvb.2011.02.001

Widmer, P. S., Semmer, N. K., Kälin, W., Jacobshagen, N., & Meier, L. L. (2012). The ambivalence of challenge stressors: Time pressure associated with both negative and positive well-being. *Journal of Vocational Behavior, 80*(2), 422–433. doi:10.1016/j.jvb.2011.09.006

Author Biographies

Matea Paškvan received her Ph.D. in work and organizational psychology from the University of Vienna. She currently works at the analysis and projections department of Statistics Austria. Her research focuses on the effects of intensified job demands on employee well-being and motivation as well as on the role of cognitive appraisal in stress processes.

Bettina Kubicek is Assistant Professor of Work and Organizational Psychology at the University of Vienna, Austria. Her current research focuses on new demands, especially on the positive and negative effects of work intensification and flexible and boundaryless work on employee well-being and the work-home interface. She is also interested in emotion work, older workers and retirement.

Chapter 4
The Bright and Dark Sides of Job Autonomy

Bettina Kubicek, Matea Paškvan and Johanna Bunner

Over the past few decades, the regulation of work has to an increasing extent been handed over to individual workers, who are given greater autonomy in performing their jobs (Allvin et al. 2011; see also Flecker et al., Chap. 2). Employees now have greater discretion over when, where, and with what means they pursue their work tasks (Wood 2011). Although an array of theoretical and empirical accounts has praised the positive effects of job autonomy for employee well-being and motivation since the 1970s (e.g., Karasek 1979), a small body of empirical work indicates that too much job autonomy may be detrimental to employees (Warr 2013). Bringing together these diverse research streams, this chapter aims to shed light on the costs and benefits of job autonomy in today's working world. It discusses the different meanings of job autonomy in highly regulated and more flexible forms of work and offers insights on when job autonomy has beneficial and deleterious effects. Furthermore, it proposes that the beneficial or deleterious effects of autonomy are contingent on characteristics of the individual worker as well as on constellations of job characteristics.

B. Kubicek (✉) · M. Paškvan · J. Bunner
Faculty of Psychology, Universitätsstraße 7, 1010 Vienna, Austria
e-mail: bettina.kubicek@univie.ac.at

M. Paškvan
e-mail: matea.paskvan@univie.ac.at

J. Bunner
e-mail: johanna.bunner@univie.ac.at

C. Korunka and B. Kubicek (eds.), *Job Demands in a Changing World of Work*,
DOI 10.1007/978-3-319-54678-0_4

4.1 Setting the Stage: The Definition of Job Autonomy

Initially, job autonomy (which is also referred to as decision latitude or job control, e.g., Karasek 1979) was conceptualized as the extent to which individuals can freely choose how to carry out their work tasks (e.g., Hackman and Oldham 1976). Recently, this broad conceptualization has been split up into distinct aspects of job autonomy and has been supplemented by additional facets. For example, Morgeson and Humphrey (2006) distinguish between discretion in scheduling work tasks (scheduling autonomy), making task-related decisions (planning autonomy), and selecting work methods (methods autonomy) as components of job autonomy. While these aspects of job autonomy are situated at the task level of a job, recent developments such as the expansion of flexible work arrangements have brought about autonomy at the job level. Employees now also have autonomy in deciding when (working time autonomy) and where (workplace autonomy) they perform their jobs (Ala-Mursula et al. 2005; Nijp et al. 2012, see also Gerdenitsch, Chap. 6). Bringing these different facets together, job autonomy refers to the discretion employees have over when, where, in which order and with what means they pursue their work tasks.

4.2 The Bright Side of Job Autonomy

According to the Job Demand-Control model (Karasek 1979) and the more general Job Demands-Resources model (Bakker and Demerouti 2007; Demerouti et al. 2001), autonomy is regarded as a job resource, which has exclusively positive effects on work-related outcomes such as well-being, job satisfaction and performance.

The *Job Demand-Control model* (Karasek 1979) posits that jobs are characterized by two important features, namely job demands and job control. Whereas job demands typically refer to quantitative demands in terms of workload or time pressure, job control or autonomy reflects the discretion employees have over deciding how to pursue their tasks. The Job Demand-Control model follows two basic assumptions: first, strain increases as demands increase. Second, a growth in competency can be expected when individuals are equipped to deal with demanding jobs, that is, when they have discretion in deciding how to approach job demands. Karasek (1979) postulates four types of jobs which result from different combinations of job demands and job decision latitude (see Fig. 4.1 for an illustration).

When job demands are high but job control is low, mental strain can be expected. These so-called "high strain jobs" lead to physical distress and psychological stress in terms of reduced well-being and even a temporary inability to work (Karasek and Theorell 1990). In contrast, when job demands are high but job control is simultaneously high, the job is considered to be active and motivates employees to learn and develop new behavioral patterns. Passive jobs, on the other

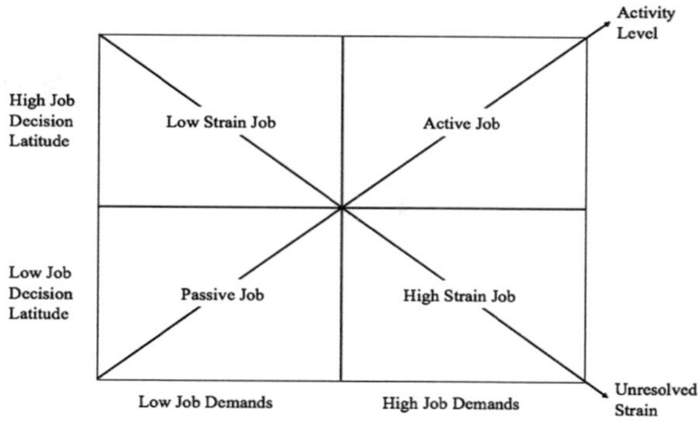

Fig. 4.1 The Job Demand-Control model (adapted from Karasek 1979)

hand, are characterized by low job demands and low job control and lead to inactivity and a decrease in general problem solving activity (Suomi and Harlow 1972). Finally, low strain jobs are characterized by low levels of job demands and high levels of job control.

The *Job Demands-Resources model* (Demerouti et al. 2001) may be seen as a generalization of the Job Demand-Control model. A central assumption of the Job Demands-Resources model is that every job has its own characteristics, which are separable into two main categories – job demands and job resources (Bakker and Demerouti 2007; Bakker et al. 2005; Demerouti et al. 2001). Job demands are defined as physical, psychological, social or organizational aspects of the job that require constant physical or psychological, cognitive or emotional effort and that are associated with physiological and psychological costs respectively (Bakker and Demerouti 2007). Examples of job demands are high workload, time pressure, unfavorable physical environment (e.g., noise, heat) and emotionally demanding interactions with clients. Job resources, on the other hand, are understood as physical, psychological, social or organizational aspects that are characterized by some or all of the following features: first, they functionally help people achieve work goals; second, they minimize certain job demands and the corresponding costs; third, they stimulate personal growth, learning and progress (Demerouti et al. 2001). According to this definition, job resources not only serve a function in terms of dealing with job demands, but are of importance in and of themselves. They can either be intrinsically motivating, if they support personal growth, or extrinsically motivating, if they are seen as a means to reach certain work-related goals (Bakker and Demerouti 2007). Therefore, gaining and conserving resources not only buffers the detrimental effects of job demands but also results in higher work engagement, less cynicism and better performance. Job resources can be found at different levels – the organization (e.g., pay, career opportunities, job security), social relations (e.g., support from supervisors and colleagues, team climate), work organization (e.g., role clarity, participation in decision making), and tasks (e.g., job autonomy, skill and task variety, task significance, performance feedback). According to the Job Demands-Resources model, job autonomy represents one of many job resources situated at the task level.

Following from the Job Demand-Control model and the Job Demands-Resources model, job autonomy has a dual role. It helps employees reach work goals and fosters motivation as well as personal growth. Additionally, it counteracts the physiological or psychological costs of job demands. Supporting the first assumption that job autonomy is of value in and of itself, meta-analytical findings have confirmed that job autonomy is positively related to the acquisition of knowledge and skills and to workers' motivation for learning (Wielenga-Meijer et al. 2010) as well as to job motivation, work engagement, and performance (Ng and Feldman 2015). Moreover, there is cumulative evidence that job autonomy not only allows employees to achieve their work goals in terms of productivity but also in terms of safety. The more autonomy employees have to carry out their work, the fewer adverse events (near misses, safety events, and errors), accidents, injuries, and unsafe behaviors occur (Nahrgang et al. 2011). Finally, job autonomy reduces conflicts arising from incompatible role pressures from the work and family domains (Michel et al. 2011; Ng and Feldman 2015). That is, the more autonomous employees are in their jobs, the less work hinders them from performing their family roles.

However, results regarding the importance of job autonomy in counteracting the adverse effects of job demands are not that unequivocal. Although there is strong evidence that higher levels of job autonomy are associated with less job stress and burnout, better mental and physical health (Ng and Feldman 2015; Park et al. 2014; Rau and Buyken 2015), as well as lower risk for cardiovascular diseases (Fishta and Backé 2015) and sleep problems (Linton et al. 2015), there is only weak empirical support for the assumption that job autonomy does indeed counteract the negative effects of high demands on employee well-being and physical health (van der Doef and Maes 1998, 1999; Häusser et al. 2010). Given this weak empirical support, Taris (2006) raised the question of whether the buffer hypothesis is a "zombie theory" which should long ago have died from a lack of empirical evidence but is kept alive in theoretical debates and empirical research. On the other hand, the findings by Häusser et al. (2010) suggest that we should not rashly abandon the buffer hypothesis. Although the buffer hypothesis has been supported by only a minority of studies, the direction of the statistically significant buffering effects have generally been in alignment with the theoretical predictions of the Job Demand-Control model (Karasek 1979). Additionally, support for the buffer hypothesis was found more often when job demand and job control measures belonged to the same level of functioning (Häussler et al. 2010). That is, scheduling autonomy was found to be more likely to mitigate the negative effects of time pressure on employee well-being than the effects of emotional demands. Therefore, Häussler et al. (2010) conclude that buffering effects do exist and that the likelihood of detecting them in empirical research depends upon the degree to which demands and control match. This match idea has recently been extended by De Jonge and Dormann (2006), who also included different types of strain. Specifically, the triple match principle states that an interaction is most likely to occur when demands, resources, and strains are based on qualitatively similar domains. To illustrate, scheduling control is most likely to mitigate the negative effects of time pressure on exhaustion. Preliminary support for the triple match principle comes from two longitudinal studies (De Jonge and Dormann 2006) showing that the likelihood of finding buffering effects was linearly related to the degree of match between demands, control, and strains.

According to the findings reviewed above, it seems that job autonomy has positive effects on employee well-being and has the potential to buffer the negative effects of job demands. However, research insights also suggest that autonomy might not be an exclusively valuable resource.

4.3 The Dark Side of Job Autonomy?

In contrast to the Job Demand-Control model (Karasek 1979) and the Job Demands-Resources model (Bakker and Demerouti 2007; Demerouti et al. 2001), the Vitamin model (Warr 1990, 1994) challenges the assumption that job autonomy is generally beneficial. The model proposes that the effects of job characteristics on mental health are analogous to the effects of vitamins on physical health. Whereas some vitamins are generally positive regardless of the extent to which they are consumed (i.e., vitamins C and E), others become detrimental to people's health with excessive consumption (i.e., vitamins A and D). More precisely, Warr (1994: 88) states that "the intake of vitamins is important for physical health up to, but not beyond, a certain level" and that excessive vitamin consumption can even be harmful to a person's health. Certain features of the work environment are proposed to have similar properties as vitamins, such that after a certain point, an increase in beneficial job characteristics may no longer have a beneficial effect on employees (i.e., constant effect) or may even be detrimental (i.e., additional decrement; for an illustration of these effects, see Fig. 4.2). Among the job characteristics that are assumed to have a constant effect are wage and status. Job characteristics assumed to suffer from additional decrement include job autonomy, skill use, and interpersonal contact.

Warr (1994) provides three different explanations of the additional decrement effect of job autonomy: one intrinsic and two additional explanatory mechanisms. The intrinsic mechanism explains how the nature of high levels of job autonomy

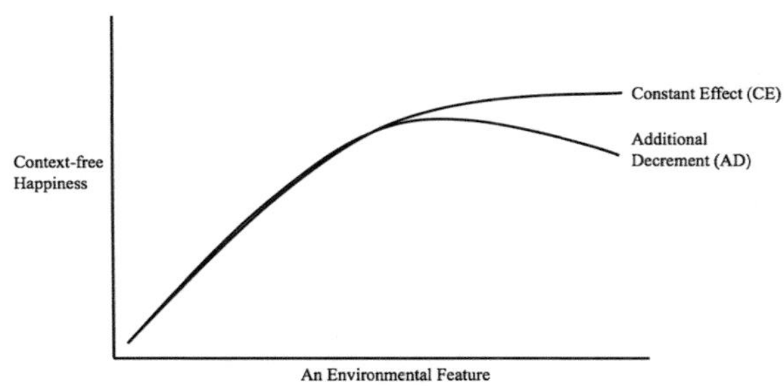

Fig. 4.2 Constant effects and additional decrement effects of job characteristics on employee well-being (adapted from Warr 2013)

contributes to detrimental effects on employee well-being. According to this explanation, high levels of autonomy may no longer be considered a "nicety" but more a workplace "necessity" i.e., inhibiting action rather than fostering it. The additional explanatory mechanisms refer to constellations of job characteristics which may contribute to the deleterious effects of job autonomy. On the one hand, high levels of autonomy may be associated with high levels of several undesirable job characteristics such as high time pressure. On the other hand, high levels of autonomy may be associated with lower levels of other desirable job characteristics such as workplace support.

Several studies have tested the assumptions of the Vitamin model, especially focusing on autonomy and different aspects of work-related well-being (e.g., burnout, job satisfaction). Warr (1990), for instance, found support for the propositions made by the Vitamin model in a heterogeneous sample. He found that job autonomy had a non-linear relation with job satisfaction. However, instead of resembling an inverted U-shape, the relation was concave downward with smaller increases in job satisfaction seen at higher levels of autonomy. Full support for the assumptions of the Vitamin model (Warr 1990, 1994) was found in a cross-sectional and a longitudinal study of eldercare workers (Kubicek et al. 2014). Eldercare workers with high and low levels of autonomy reported a higher tendency to be irritated (cross-sectionally) and to depersonalize care recipients as well as a lower tendency to feel dedicated to, absorbed by and full of energy at work (longitudinally) than those with medium levels of autonomy. Likewise, Wieland et al. (2004) found negative effects of high job autonomy in a sample of IT workers. Specifically, high job autonomy led to increased task insecurity, which in turn was related to burnout symptoms. With regard to job satisfaction, Baltes et al. (2002) found that job autonomy follows a curvilinear function: job satisfaction increased when job autonomy increased from low to medium levels, but it decreased again at very high levels of job autonomy. Conversely, in a sample of civil servants, cross-sectional as well as longitudinal evidence indicated a J-shaped relation between autonomy and job satisfaction, where higher levels of autonomy were associated with higher job satisfaction (Rydstedt et al. 2006). In a study among health care workers, the effect of autonomy was again not in line with the assumptions of the Vitamin model (De Jonge and Schaufeli 1998). Instead of a U-shaped curve, an inverted U-shaped curve was found, where low and high levels of autonomy were related to lower levels of exhaustion. Moreover, several studies failed to find non-linear associations between autonomy and work-related outcomes (Jeurissen and Nyklíček 2010; Mäkikangas et al. 2007). Collectively, previous studies provide some support for non-linear relations between autonomy and aspects of work-related well-being. In general, however, the results are inconsistent.

These inconsistencies led Taris (2006) to generally question the existence of curvilinear relations between job characteristics such as job autonomy and employee well-being. He even labeled the notion of curvilinearity an "urban myth" within work and organizational psychology. Yet there is still reason to believe that high levels of autonomy can be detrimental for employee well-being and that the weak support for curvilinear relations is partly due to methodological difficulties in detecting nonlinear relations. One such methodological problem is the prevalence of mono-method research designs, where job autonomy and its outcomes are measured using the same format. Baltes et al. (2002), for instance, showed that the nonlinear effects of job autonomy contributed more strongly to the prediction of job satisfaction when using multiple methods to assess job autonomy and its work-related outcomes than in the case of a single measurement method.

4.4 The Meaning of Job Autonomy in Regulated and Flexible Forms of Work

Nowadays, the direct "Tayloristic" method of controlling employees' labor seems to be outdated. To promote innovation and customized products and services in globalized "economies of speed", organizations are reducing direct control and increasing requirements for autonomous planning and decision-making on the part of employees. However, having more autonomy may mean qualitatively different things in work settings that are highly flexible and characterized by indirect control than in work settings that are highly regulated and externally controlled (such as manual work in the 1980s). While an increase in autonomy may be generally beneficial in highly regulated and externally controlled work settings, it may be accompanied with more uncertainty and ambiguity (see e.g., Johlke and Iyer 2013) in highly flexible and indirectly controlled work settings. In these work settings, employees not only have the possibility of making decisions on their own; rather, they are forced to do so. They need to act as flexible entrepreneurs (Pongratz and Voß 2003) who are increasingly expected to plan and structure their workday autonomously, determine how to handle work tasks as well as autonomously control work goals (Kubicek et al. 2015). Under these circumstances, job autonomy may no longer be a nicety but rather a necessity that requires additional effort and hinders task completion (Warr 2013). In line with this argument, a study analyzing longitudinal effects of intensified planning and decision-making demands showed that eldercare workers who reported an intensification of planning and decision-making were more cognitively irritated after regular work hours and reported higher levels of emotional exhaustion two years later (Paškvan et al. 2013). This study highlights that the increasing need to control, structure, and plan work is a demand which can lead to strain. This might especially be the case in flexible work arrangements where autonomous decision-making seems to be an integral part of the job (see Gerdenitsch, Chap. 6).

Additional support for the assumption that job autonomy may lose its beneficial effects in highly flexible work settings can be found in the literature on worktime and workplace autonomy. In their meta-analysis of the effects of worktime autonomy (i.e., flexible and compressed workweek schedules) on work-related outcomes, Baltes et al. (1999) found that highly flexible work schedules had diminished effectiveness in comparison with less flexible programs. They argue that high levels of flexibility may, for example, be an inconvenience to employees who are highly dependent on their colleagues for task completion. Problems with scheduling may adversely affect communication, supervision and task performance in cases where worktime autonomy is especially high.

Moreover, research on knowledge workers reveals a tension between autonomy and control that has led to the formulation of the so-called "autonomy paradox" (Mazmanian et al. 2013). This paradox posits that "the more autonomy employees have, the harder they work, the more hours they devote, and the more organizations control their lives" (Putnam et al. 2014: 427). The autonomy paradox also becomes visible in studies on worktime and workplace autonomy. To illustrate, employees who have autonomy in their schedules, times, and places of work often intensify their work effort (Kelliher and Anderson 2010; see also Paškvan and Kubicek, Chap. 3). Sennett (1998) describes this phenomenon as the corrosive effect of flexible work: Rather than decreasing organizational control, autonomy in work settings often increases control. In these situations, the job exerts control despite the autonomy employees have via their flexible work schedules and work places. Norms and values of increased work effort (such as a long-hours culture, Chatzitheochari and Arber 2009), the need to maintain or establish professional reputation (Gerdenitsch et al. 2015), and externally-set deadlines influence and regulate employees who have autonomy over their own hours and schedules. At the same time, autonomy disguises the fact that these norms are forms of organizational control. Consequently, workers with worktime and workplace autonomy are not aware that they are controlled by norms, values, and socialization practices or images of the ideal worker (Putnam et al. 2014). This paradoxical nature of high levels of workplace and worktime autonomy may explain why such forms of autonomy are not positively related to employee well-being and health (Nijp et al. 2012), while task-related autonomy has been consistently shown to reduce job stress and increase well-being. Given the potentially bright and dark sides of autonomy, the question arises as to when job autonomy has beneficial or detrimental effects on employee well-being.

4.5 Beneficial or Deleterious Effects of Job Autonomy? What Makes the Difference?

The trend of giving employees greater autonomy (Wood 2011) requires individual workers to take increased responsibility for their work. Hence, the worker's predispositions and abilities gain importance for the successful completion of work (Warr 2013). Whether or not workers can handle this greater degree of discretion in performing their work is therefore partially contingent on individual factors (Van Yperen et al. 2014).

4.5.1 Employees' Personal Characteristics

An important personal characteristic which determines how positively or negatively employees react to job autonomy is their *need for autonomy*. According to self-determination theory (Gagné and Deci 2005), people differ in three central needs – the need for autonomy, competence, and relatedness. Additionally, self-determination theory postulates that motivation exists on a continuum of controlled versus autonomous motivation, and internalization is the process leading from controlled (external) to autonomous (intrinsic) motivation. People who are autonomously motivated *want* to act and choose to do so on their own (Gagné and Deci 2005). Self-determination theory further assumes that environments differ in the extent to which they provide support for individuals' needs for autonomy, competence and relatedness and that the satisfaction of these psychological needs forms the foundation for autonomous motivation and internalization. In order to internalize a certain behavior, it is necessary for an individual to experience a feeling of competence and relatedness with regard to that behavior. The degree to which the need for autonomy can be satisfied, then, determines to which extent the behavior is internalized (whether it is merely introjected, or if it can be identified with or even integrated). The need for autonomy in particular is suggested to be the most important predictor of autonomous (or intrinsic) motivation, and several studies support this claim. Lynch et al. (2005) conducted a study among psychiatric hospital staff and found evidence that supervisors who supported greater autonomy among their subordinates fostered increased internalization compared to supervisors who were perceived as controlling. Moreover, studies among managers have shown that behavior supportive of autonomy leads to increased fulfillment of the needs for autonomy, competence and relatedness and in turn results in higher levels of trust and positive work-related attitudes (Deci et al. 1989), increased job satisfaction, higher performance evaluations, greater persistence, greater acceptance of organizational change, and better psychological adjustment (Baard et al. 2004; Deci et al. 2001; Gagné et al. 2000; Ilardi et al. 1993; Kasser et al. 1992).

Apart from need satisfaction, people differ in the extent to which they have psychological needs. Therefore, the effects of job autonomy on its outcomes are likely to differ between people with a high need for autonomy and those with a low need for autonomy. For example, Van Yperen et al. (2014) showed that worktime and workplace autonomy had the most positive effects on employees with a high need for autonomy. As the opportunity to choose the times and places of work increases employees' autonomy, employees with a high need for autonomy may feel especially fulfilled in such working conditions. Further supporting this assumption, Van Yperen et al. (2016, in press) showed that even simply perceiving an opportunity to decide when and where to work (independent of the actual use of this opportunity) acted as an effective resource to cope with increasing job demands for employees with a high need for autonomy, but not for employees with a low need for autonomy. Likewise, in a study on the interactive effects of job autonomy and the need for autonomy, Gaziel (1989) found that job autonomy was not related to motivation and job satisfaction among administrators with a low need for autonomy, whereas a positive relation between job autonomy and the outcome measures was found for administrators with a high need for autonomy. Gaziel (1989) concluded that a lack of job autonomy is experienced if and only if there is a discrepancy between the actual level and the desired level of autonomy. In a similar vein, Park and Jang (2015) found in a nationally representative survey that the beneficial effects of job autonomy were contingent on employees' individual work values. While they found positive links between job autonomy and mental health, this link was stronger among employees with strong intrinsic work values than among employees with strong extrinsic work values. Thus, when designing work environments, supervisors should acknowledge individual differences in their employees' need for autonomy and work values.

In addition to employees' need for autonomy, their level of *self-efficacy* may influence whether job autonomy has positive or negative effects. Self-efficacy is understood as the "beliefs in one's capabilities to organize and execute the courses of action required to produce given attainments" (Bandura 1997: 3). The self-efficacious individual therefore believes him/herself capable of managing context-specific demands (Chen et al. 2001). Furthermore, self-efficacious individuals seem to prefer different types of jobs and work environments compared to individuals who consider themselves inefficacious (Gibson 2001; Jex and Bliese 1999), and different expressions of self-efficacy impact the extent to which employees prefer responsibility and challenges in their jobs (Bandura 1997; Jex and Bliese 1999). Highly efficacious employees aspire for jobs with a "high scope", appreciate more challenging jobs, and do not back away anxiously when faced with threatening tasks or environments but rather intensify their efforts to make their performance match the demands placed on them (Bandura 1986; Jex and Bliese 1999). In contrast, individuals who consider themselves to be inefficacious are prone to avoiding difficult challenges or tasks, tend to give up easily and appear more anxious (Bandura 1986). Therefore, it seems likely that employees' self-efficacy influences how they react to job autonomy. In a study of 733 health care employees, van Mierlo et al. (2006) showed that the more self-efficacious team members were, the more individual autonomy they reported. Additionally, Salanova et al. (2002) showed that levels of specific self-efficacy (i.e., computer self-efficacy) moderated the relationship between job demands and control and levels of burnout dimensions. The most relevant result was that high levels of computer self-efficacy buffered burnout when job demands and job control were high. Moreover, for workers with high computer self-efficacy but low control at work, high demands increased levels of burnout, as did they for workers with low computer self-efficacy but high control at work. These results are in alignment with Jimmieson (2000) and Schaubroeck and Merritt (1997), who found that Karasek's model was supported among people who experienced a high sense of self-efficacy. Among people with low job self-efficacy, increasing control at work may exacerbate the stress of demanding jobs, while high self-efficacy may likewise be harmful when people have no control at work. In this case, people may become discouraged when their self-efficacious beliefs are challenged by the objective uncontrollability of a situation (Wortman and Dunkel-Schetter 1979; cf., Schaubroeck and Merritt 1997).

Finally, research comparing younger and older persons' work attitudes suggests that today's younger generation values autonomy and freedom in their lives more than prior generations (Twenge 2010). *Workers' age* may thus be an important factor influencing the relations between job autonomy and work-related as well as work-home outcomes. Socio-emotional selectivity theory (Carstensen 1991) proposes that different age groups prefer different kinds of social activities and emotional experiences. According to Carstensen (1991), adults have two main goals in life: knowledge acquisition and emotional regulation. Knowledge acquisition entails behaviors directed at learning new knowledge and acquiring information as well as using that knowledge and information to build one's career. Emotional regulation, on the other hand, entails behaviors directed towards finding meaning in life, developing close relationships with others and developing a sense of belonging. Moreover, Carstensen (1991) proposes and shows empirically that the relative importance of these activities changes over the life span. While younger adults engage more strongly in activities directed towards the acquisition of knowledge and information which help them advance their careers, older adults prefer to engage in social activities which provide support and affirmation. Thus, according to socio-emotional selectivity theory (Carstensen 1991), younger workers may react more positively to job autonomy because it allows them to acquire knowledge and information and to more strongly develop their job skills and demonstrate their work competence. In line with this is a longitudinal study that found no relation between intensified job- and career-related autonomy demands and irritation among young professionals over a period of six months (Bunner 2014). This is in agreement with self-determination theory, which posits that individuals react especially positively to work environments that fulfill their basic human needs (Gagné and Deci 2005). Empirical evidence partly supports this assumption. In a meta-analysis on the moderating effect of age on the relations between job autonomy and a wide range of work-related and work-home outcomes, Ng and Feldman (2015) showed that job autonomy was less strongly related to job satisfaction, affective commitment, work engagement, job stress, and poor mental health for older workers than for younger workers. Thus, younger workers indeed seem to react more positively to job autonomy than older workers. However, Ng and Feldman (2015) also found a stronger relation between job autonomy and job performance as well as (less) emotional exhaustion among older workers.

4.5.2 Constellations of Job Characteristics

According to Warr (2013), high levels of job autonomy become deleterious for employee well-being if they are associated with high levels of undesirable job characteristics such as high time pressure or lower levels of desirable job characteristics such as workplace support. Therefore, it seems important to also take constellations of job characteristics into account when assessing the conditions under which job autonomy is associated with potential costs or benefits.

Now, why should high levels of autonomy become harmful when they are accompanied by high levels of *time pressure*? With regard to employee well-being, considerable autonomy goes along with ambiguity (see e.g., Johlke and Iyer 2013) and substantial responsibility and may therefore make decision-making more difficult. Extensive decision-making demands are likely to require cognitive regulation at the highest intellectual level (Hacker 2003). When regulating work tasks at this level of action regulation, employees expend an increased amount of cognitive effort, which comes with psychological costs, such as the depletion of cognitive resources (Baumeister et al. 1998). Being concurrently exposed to time pressure may make it impossible for employees to replenish their depleted cognitive resources, contributing to the adverse effects of high levels of autonomy on employee well-being and increasing the likelihood of experiencing job stress (Paškvan et al., in prep). In addition to impairing employees' well-being by contributing to job stress, such constellations may also hamper employee work motivation. High levels of autonomy and time pressure may give rise to work anxiety (Warr 1990) or beliefs that one is not able to successfully conform to goal expectations (Preston 2013). Therefore, high autonomy in conjunction with high time pressure may be appraised as threatening and may thus interfere with feelings of intrinsic motivation (Gagné and Deci 2005).

Time pressure, however, does not seem to be the only job characteristic that determines whether job autonomy is beneficial or detrimental to employees' well-being, work behavior, and work attitudes. *Job complexity* is yet another job characteristic that interacts with autonomy in predicting work-related outcomes. Job complexity refers to the extent to which jobs are mentally challenging and require workers to use complex skills (e.g., Morgeson and Humphrey 2006). Complex jobs require considering different goals and plans (e.g., Frese 1987), solving unforeseen problems, and dealing with unpredictable events (Kohn and Schooler 1983). Therefore, complex jobs are also characterized by ambiguity and a lack of structure. According to activation theory (Gardner 1986), optimal levels of activation or stimulation are beneficial for performance. Too little activation leads to boredom, while too much activation results in mental overload. Thus, although job autonomy in conjunction with job complexity should provide greater stimulation to employees, very high levels of job autonomy and job complexity may result in too much activation, impairing performance, satisfaction, and well-being. Supporting this assumption, Chung-Yan (2010) showed that job autonomy was curvilinearly related to job satisfaction only for employees with high job complexity, but not for those with low or moderate job complexity. More precisely, job autonomy was positively related to job satisfaction only up to a certain level. Once this threshold was exceeded, the initially positive association reversed and an increase in autonomy was associated with a decrease in job satisfaction. With regard to psychological well-being, job autonomy had more of a constant effect than an additional decrement effect among employees high in job complexity: at moderate levels of job autonomy, its positive effect on psychological well-being reached a plateau. Thereafter, an increase in job autonomy was associated with no further improvement in psychological well-being.

In addition to the conditional effects of time pressure and job complexity, job autonomy may be detrimental because it is associated with lower levels of desirable job characteristics such as *workplace support* (Warr 2013). Although this is not the case for job autonomy in general, as it is shown to be mostly positively related to supervisor support (Kalleberg et al. 2009) and supervisor satisfaction (Humphrey et al. 2007), the aforementioned detrimental effect should be especially important when employees have high workplace autonomy. Being able to choose where to work may imply that employees are no longer present at a fixed time, at a fixed place and that real – in contrast to virtual – "facetime" between leaders, co-workers and employees is reduced (Kossek and Van Dyne 2011). Hence, workplace autonomy might lead to a situation where employees, co-workers, and leaders do not automatically meet each other in person but rather via electronic communication systems (Sarbaugh-Thompson and Feldmann 1998). In such a situation, the opportunities to give and receive support are greatly reduced (Demerouti et al. 2014), which in turn demands that employees actively ask for advice. These assumptions were partly confirmed by a meta-analysis showing that high-intensity telecommuting does not amplify negative effects on the supervisor relationship, but rather on the co-worker relationship (Gajendran and Harrison 2007). Hence, it can be argued that employees who take intensive advantage of workplace autonomy meet the challenge of actively asking for direct supervisor support but actually have fewer interactions with co-workers and therefore receive less support from colleagues.

4.6 Conclusion

In this chapter, we have explored the potentially positive and negative effects of job autonomy for employee well-being as well as work-related (e.g., performance) and work-home outcomes (e.g., work-home conflict). We summarized literature showing that job autonomy in terms of how individuals perform their work tasks (i.e., method and scheduling autonomy) is mostly beneficial for employees. This is underscored by a large and diverse body of empirical findings, highlighting the positive effects of job autonomy on well-being, performance, and work-home balance. However, there is also a small body of research showing that job autonomy may be detrimental in some situations. This is especially the case when job demands are high (i.e., time pressure, job complexity) or when individuals have a lower need for autonomy (be it because they value autonomy less or because of their age). Although these findings point towards potentially negative effects of high levels of autonomy for employee well-being, stronger evidence for potentially detrimental effects of autonomy can be found in studies on worktime and workplace autonomy. Employees who have autonomy in their times and places of work often intensify their work effort (Kelliher and Anderson 2010; see also Paškvan and Kubicek, Chap. 3) due to norms, career aspirations, or externally set goals. Consequently, despite ostensible autonomy through flexible work schedules, employees remain highly controlled by their jobs. In such highly flexible work settings, autonomy may become an obligation (Gerdenitsch et al. 2015) that lacks the positive effects of having decision latitude. The ambiguity and uncertainty which often accompany such highly flexible situations may also contribute to the demanding and thus impairing effects of autonomy.

From a practical point of view, the existing literature suggests that job autonomy is an important job resource that contributes to employee well-being, motivation and work-home balance. Its potential to counteract the negative effects of job demands is, however, limited and depends on the match between the demands at hand and the form of control given to employees. Moreover, the positive effects of autonomy are contingent on the form of autonomy that is employed (method and scheduling autonomy versus worktime and workplace autonomy) as well as on constellations of job characteristics and individual characteristics. Thus, there is no best way of implementing job autonomy, as high levels of autonomy may in some cases mean having too much of a good thing (Pierce and Aguinis 2013).

References

Ala-Mursula, L., Vahtera, J., Linna, A., Pentti, J., & Kivimäki, M. (2005). Employee worktime control moderates the effects of job strain and effort-reward imbalance on sickness absence: The 10-town study. *Journal of Epidemiology and Community Health, 59,* 851–857.

Allvin, M., Aronsson, G., Hagström, T., Johansson, G., & Lundberg, U. (2011). *Work without boundaries. Psychological perspectives on the new working life*. Chichester, UK: Wiley-Blackwell.

Baard, P. P., Deci, E. L., & Ryan, R. M. (2004). The relation of intrinsic need satisfaction to performance and wellbeing in two work settings. *Journal of Applied Social Psychology, 34,* 2045–2068.

Bakker, A. B., & Demerouti, E. (2007). The job demands-resources model: State of the art. *Journal of Managerial Psychology, 22*, 309–328. doi:10.1108/02683940710733115.

Bakker, A. B., Demerouti, E., & Euwema, M. C. (2005). Job resources may buffer the impact of job demands on burnout. *Journal of Occupational Health Psychology, 10,* 170–180.

Baltes, B. B., Bauer, C. C., Bajdo, L. M., & Parker, C. P. (2002). The use of multitrait-multimethod data for detecting nonlinear relationships: The case of psychological climate and job satisfaction. *Journal of Business and Psychology, 17,* 3–17.

Baltes, B. B., Briggs, T. E., Huff, J. W., Wright, J. A., & Neuman, G. A. (1999). Flexible and compressed workweek schedules: A meta-analysis of their effects on work-related criteria. *Journal of Applied Psychology, 84,* 496–513.

Bandura, A. (1986). *Social foundations of thought and action: A social cognitive theory.* Englewood Cliffs, NJ: Prentice Hall.

Bandura, A. (1997). *Self-efficacy: The exercise of control*. New York: Freeman.

Baumeister, R. F., Bratslavsky, E., Muraven, M., & Tice, D. M. (1998). Ego depletion: Is the active self a limited resource? *Journal of Personality and Social Psychology, 74,* 1252–1265.

Bunner, J. (2014). How do professional newcomers deal with increased autonomy at work? The importance of self-leadership styles (Unpublished diploma thesis). Wien: Universität Wien.

Carstensen, L. L. (1991). Selectivity theory: Social activity in life-span context. *Annual Review of Gerontology and Geriatrics, 11,* 195–217.

Chatzitheochari, S., & Arber, S. (2009). Lack of sleep, work and the long hours culture: Evidence from the UK time use survey. *Work, Employment & Society, 23,* 30–48.

Chen, G., Gully, S. M., & Eden, D. (2001). Validation of a new general self-efficacy scale. *Organizational Research Methods, 4,* 62–83.

Chung-Yan, G. A. (2010). The nonlinear effects of job complexity and autonomy on job satisfaction, turnover, and psychological well-being. *Journal of Occupational Health Psychology, 15,* 237–251.

Deci, E. L., Connell, J. P., & Ryan, R. M. (1989). Self-determination in a work organization. *Journal of Applied Psychology, 74,* 580–590. http://doi.org/10.1037/0021-9010.74.4.580.

Deci, E. L., Ryan, R. M., Gagne, M., Leone, D. R., Usunov, J., & Kornazheva, B. P. (2001). Need satisfaction, motivation, and well-being in the work organizations of a former eastern bloc country: A cross-cultural study of self-determination. *Personality and Social Psychology Bulletin, 27,* 930–942. doi:10.1177/0146167201278002.

De Jonge, J., & Dormann, C. (2006). Stressors, resources, and strain at work: A longitudinal test of the triple-match principle. *Journal of Applied Psychology, 91,* 1359–1374.

De Jonge, J., & Schaufeli, W. B. (1998). Job characteristics and employee well-being: A test of Warr's vitamin model in health care workers using structural equation modelling. *Journal of Organizational Behavior, 19,* 387–407.

Demerouti, E., Bakker, A. B., Nachreiner, F., & Schaufeli, W. B. (2001). The job demands resources model of burnout. *Journal of Applied Psychology, 86,* 499–512.

Demerouti, E., Derks, D., Lieke, L., & Bakker, A. B. (2014). New ways of working: Impact on working conditions, work–family balance, and well-being. In C. Korunka & P. Hoonakker (Eds.), *The impact of ICT on quality of working life* (pp. 123–141). The Netherlands: Springer.

Fishta, A., & Backé, E.-M. (2015). Psychosocial stress at work and cardiovascular diseases: An overview of systematic reviews. *International Archives of Occupational and Environmental Health, 88,* 997–1014.

Frese, M. (1987). A theory of control and complexity: Implications for software design and integration of computer systems into the work place. In M. Frese, E. Ulich, & W. Dzida (Eds.), *Psychological issues of human computer interaction in the work place* (pp. 313–337). Amsterdam: Elsevier.

Gagné, M., & Deci, E. L. (2005). Self-determination theory and work motivation. *Journal of Organizational Behavior, 26,* 331–362. http://doi.org/10.1002/job.322.

Gagné, M., Koestner, R., & Zuckerman, M. (2000). Facilitating acceptance of organizational change: The importance of self-determination. *Journal of Applied Social Psychology, 30,* 1843–1852.

Gardner, D. G. (1986). Activation and theory and task design: An empirical test of several new predictions. *Journal of Applied Psychology, 71,* 411–418.

Gaziel, H. (1989). Determinants of perceived deficiency of autonomy among elementary school administrators. *Social Behavior and Personality, 17,* 57–65.

Gajendran, R. S., & Harrison, D. A. (2007). The good, the bad, and the unknown about telecommuting: meta-analysis of psychological mediators and individual consequences. *Journal of Applied Psychology, 92,* 1524.

Gerdenitsch, C., Kubicek, B., & Korunka, C. (2015). Control in flexible work arrangements: When freedom becomes duty. *Journal of Personnel Psychology, 14,* 61–69.

Gibson, C. B. (2001). Me and us: Differential relationships among goal-setting training, efficacy, and effectiveness at the individual and team level. *Journal of Organizational Behavior, 22,* 789–808.

Hacker, W. (2003). Action regulation theory: A practical tool for the design of modern work processes? *European Journal of Work and Organizational Psychology, 12,* 105–130.

Hackman, J. R., & Oldham, G. R. (1976). Motivation through the design of work: Test of a theory. *Organizational Behavior and Human Performance, 16,* 250–279.

Häusser, J. A., Mojzisch, A., Niesel, M., & Schulz-Hardt, S. (2010). Ten years on: A review of recent research on the Job Demand-Control (-Support) model and psychological well-being. *Work & Stress, 24,* 1–35.

Humphrey, S. E., Nahrgang, J. D., & Morgeson, F. P. (2007). Integrating motivational, social, and contextual work design features: A meta-analytic summary and theoretical extension of the work design literature. *Journal of Applied Psychology, 92,* 1332–1356.

Ilardi, B. C., Leone, D., Kasser, T., & Ryan, R. M. (1993). Employee and supervisor ratings of motivation: Main effects and discrepancies associated with job satisfaction and adjustment in a factory setting. *Journal of Applied Social Psychology, 23,* 1789–1805.

Jeurissen, T., & Nyklíček, I. (2010). Testing the vitamin model of job stress in Dutch health care workers. *Work & Stress, 15*, 254–264. doi:10.1080/02678370110066607.

Jex, S. M., & Bliese, P. D. (1999). Efficacy beliefs as a moderator of the impact of work-related stressors: A multilevel study. *Journal of Applied Psychology, 84,* 349–361.

Jimmieson, N. L. (2000). Employee reactions to behavioural control under conditions of stress: The moderating role of self-efficacy. *Work & Stress, 14,* 262–280.

Johlke, M. C., & Iyer, R. (2013). A model of retail job characteristics, employee role ambiguity, external customer mind-set, and sales performance. *Journal of Retailing and Consumer Services, 20*(1), 58–67.

Kalleberg, A. L., Nesheim, T., & Olsen, K. M. (2009). Is participation good or bad for workers? Effects of autonomy, consultation and teamwork on stress among workers in Norway. *Acta Sociologica*, *52*, 99–116.

Karasek, R. A. (1979). Job demands, job decision latitude, and mental strain: Implications for job redesign. *Administrative Science Quarterly, 24,* 285–308.

Karasek, R. A., & Theorell, T. (1990). *Healthy work*. New York: Basic Books.

Kasser, T., Davey, J., & Ryan, R. M. (1992). Motivation and employee-supervisor discrepancies in a psychiatric vocational rehabilitation setting. *Rehabilitation Psychology, 37,* 175–187.

Kelliher, C., & Anderson, D. (2010). Doing more with less? Flexible work practices and the intensification of work. *Human Relations, 63,* 83–106.

Kohn, M. L., & Schooler, C. (Eds.). (1983). *Work and personality*. Norwood, NJ: Ablex Publishing.

Kossek, E. E., & Van Dynet, L. (2011). How work-life flexibility influences work performance of individuals and groups. In K. Korabik, D. S. Lero, & D. L. Whitehead (Eds.), *Handbook of work-family integration: Research, theory, and best practices* (pp. 305–330). London: Elsevier.

Kubicek, B., Korunka, C., & Tement, S. (2014). Too much job control? Two studies on curvilinear relations between job control and eldercare workers' well-being. *International Journal of Nursing Studies, 51,* 1644–1653.

Kubicek, B., Paškvan, M., & Korunka, C. (2015). Development and validation of an instrument for assessing job demands arising from accelerated change: The Intensification of Job Demands Scale (IDS). *European Journal of Work and Organizational Psychology, 24,* 898–913. doi:10.1080/1359432X.2014.979160.

Linton, S. J., Kecklund, G., Franklin, K. A., Leissner, L. C., Sivertsen, B., Lindberg, E., … Björkelund, C. (2015). The effect of the work environment on future sleep disturbances: A systematic review. *Sleep Medicine Reviews, 23*, 10–19.

Lynch, M. F., Jr., Plant, R., & Ryan, R. M. (2005). Psychological need satisfaction, motivation, attitudes, and well-being among psychiatric hospital staff and patients. *Professional Psychology, 36,* 415–425.

Mazmanian, M., Orlikowski, W. J., & Yates, J. (2013). The autonomy paradox: The implications of mobile email devices for knowledge professionals. *Organization Science, 24,* 1337–1357.

Mäkikangas, A., Feldt, T., & Kinnunen, U. (2007). Warr's scale of job-related affective well-being: A longitudinal examination of its structure and relationship with work characteristics. *Work & Stress, 21*(3), 197–219. doi:10.1080/02678370701662151.

Michel, J. S., Kotrba, L. M., Mitchelson, J. K., Clark, M. A., & Baltes, B. B. (2011). Antecedents of work-family conflict: A meta-analytic review. *Journal of Organizational Behavior, 32,* 689–725.

Morgeson, F. P., & Humphrey, S. E. (2006). The Work Design Questionnaire (WDQ): Developing and validating a comprehensive measure for assessing job design and the nature of work. *Journal of Applied Psychology, 91,* 1321–1339.

Nahrgang, J. D., Morgeson, F. P., & Hofmann, D. A. (2011). Safety at work: A meta-analytic investigation of the link between job demands, job resources, burnout, engagement, and safety outcomes. *Journal of Applied Psychology, 96,* 71–94.

Ng, T. W. H., & Feldman, D. C. (2015). The moderating effects of age in the relationships of job autonomy to work outcomes. *Work, Aging and Retirement, 1,* 64–78.

Nijp, H. H., Beckers, D. G., Geurts, S. A., Tucker, P., & Kompier, M. A. (2012). Systematic review on the association between employee worktime control and work-non-work balance, health and well-being, and job-related outcomes. *Scandinavian Journal of Work, Environment & Health, 38,* 299–313.

Park, H. I., Jacob, A. C., Wagner, S. H., & Baiden, M. (2014). Job control and burnout: A meta-analytic test of the conservation of resources model. *Applied Psychology: An International Review, 63,* 607–642.

Park, R., & Jang, S. J. (2015). Mediating role of perceived supervisor support in the relationship between job autonomy and mental health: moderating role of value–means fit. *The International Journal of Human Resource Management*, 1–21.

Paškvan, M., Kubicek, B., Keller, A. & Korunka, C. (in prep). Curvilinear effects of autonomy: Time pressure as a boundary condition.

Paškvan, M., Kubicek, B., & Korunka, C. (2013). Can autonomy lead to emotional exhaustion? Short-term strain responses explaining negative effects of autonomy. Paper presented at the Work, Stress, & Health Conference, Los Angeles, CA, USA.

Pierce, J. R., & Aguinis, H. (2013). The too-much-of-a-good-thing-effect in management. *Journal of Management, 39,* 313–338.

Pongratz, H. J., & Voß, G. G. (2003). From employee to 'entreployee': Towards a 'self-entrepreneurial' work force? *Concepts and Transformation, 8,* 239–254.

Preston, M. S. (2013). Advancing case manager motivation in child welfare: Job control's curvilinear relationship and instrumental feedback's moderating influence. *Children and Youth Services Review, 35,* 2003–2012.

Putnam, L. L., Myers, K. K., & Gailliard, B. M. (2014). Examining the tensions in workplace flexibility and exploring options for new directions. *Human Relations, 67,* 413–440.

Rau, R., & Buyken, D. (2015). Der aktuelle Kenntnisstand über Erkrankungsrisiken durch psychische Arbeitsbelastungen. Ein systematisches Review über Metaanalysen und Reviews [Current status of knowledge about health risk from mental workload: Evidence Based on a systematic review of reviews]. *Zeitschrift für Arbeits- und Organisationspsychologie, 59,* 113–129.

Rydstedt, L. W., Ferrie, J., & Head, J. (2006). Is there support for curvilinear relationships between psychosocial work characteristics and mental well-being? Cross-sectional and long-term data from the Whitehall II study. *Work & Stress, 20*(1), 6–20.

Salanova, M., Peiró, J. M., & Schaufeli, W. B. (2002). Self-efficacy specificity and burnout among information technology workers: An extension of the job demand-control model. *European Journal of work and organizational psychology, 11*(1), 1–25.

Sarbaugh-Thompson, M., & Feldman, M. S. (1998). Electronic mail and organizational communication: Does saying "Hi" really matter? *Organization Science, 9*(6), 685–698.

Schaubroeck, J., & Merritt, D. E. (1997). Divergent effects of job control on coping with work stressors: The key role of self-efficacy. *Academy of Management Journal, 40*(3), 738–754.

Sennett, R. (1998). *The corrosion of character*. New York: W. W. Norton & Company.

Suomi, S. J., & Harlow, H. F. (1972). Depressive behavior in young monkeys subjected to vertical chamber confinement. *Journal of comparative and physiological psychology, 80*(1), 11.

Taris, T. W. (2006). Bricks without clay: On urban myths in occupational health psychology. *Work & Stress, 20,* 99–140.

Twenge, J. (2010). A review of the empirical evidence on generational differences in work attitudes. *Journal of Business and Psychology, 25,* 201–210.

van der Doef, M., & Maes, S. (1998). The job demand control (-support) model and physical health outcomes: A review of the strain and buffer hypothesis. *Psychology & Health, 13,* 909–936.

van der Doef, M., & Maes, S. (1999). The job demand-control (-support) model and psychological well-being: A review of 20 years of empirical research. *Work & Stress, 13,* 87–114.

van Mierlo, H., Rutte, C. V., Vermunt, J. K., Kompier, M. A. J., & Doorewaard, J. A. M. C. (2006). Individual autonomy in work teams: The role of team autonomy, self-efficacy, and social support. *European Journal of Work and Organizational Psychology, 15*(3), 281–299.

Van Yperen, N. W., Rietzschel, E. F., & De Jonge, K. M. M. (2014). Blended working: For whom it may (not) work. *PLoS ONE, 9*(7), e102921.

Van Yperen, N. W., Wortier, B., & De Jonge, K. M. M. (2016, in press). Workers' intrinsic work motivation when job demands are high: The role of need for autonomy and perceived opportunity for blended working. *Computers in Human Behavior*.

Warr, P. B. (1990). Decision latitude, job demands and employee well-being. *Work & Stress, 4,* 285–294. doi:10.1080/02678379008256991.

Warr, P. B. (1994). A conceptual framework for the study of work and mental health. *Work & Stress, 8,* 84–97.

Warr, P. B. (2013). Jobs and job-holders: Two sources of happiness and unhappiness. In K. Cameron & A. Caza (Eds.), *Happiness and Organizations* (pp. 733–750). Oxford University Press.

Wieland, R., Klemens, S., Scherrer, K., Timm, E., & Krajewski, J. (2004). *Moderne IT-Arbeitswelt gestalten. Anforderungen, Belastungen und Ressourcen in der IT-Branche.* [Structuring the modern IT world of work. Demands, strains and resources in the IT branch.] Techniker Krankenkasse: Hamburg.

Wielenga-Meijer, E. G. A., Taris, T. W., Kompier, M. A. J., & Wigboldus, D. H. J. (2010). From task characteristics to learning: A systematic review. *Scandinavian Journal of Psychology, 51,* 363–375.

Wood, L. A. (2011). The changing nature of jobs: a meta-analysis examining changes in job characteristics over time. Master thesis, University of Georgia, Georgia, USA. Retrieved from http://dbs.galib.uga.edu/cgi-bin/getd.cgi?userid=galileo&serverno=8&instcode=publ&_cc=1.

Wortman, C. B., & Dunkel-Schetter, C. (1979). Interpersonal relationships and cancer: A theoretical analysis. *Journal of Social Issues, 35*(1), 120–155.

Author Biographies

Bettina Kubicek is Assistant Professor of Work and Organizational Psychology at the University of Vienna and Professor of Organizational Development at the University of Applied Sciences Upper Austria. Her current research focuses on new demands, especially on the positive and negative effects of work intensification and flexible and boundaryless work on employee well-being and the work-home interface. She is also interested in emotion work, older workers and retirement.

Matea Paškvan received her Ph.D. in work and organizational psychology from the University of Vienna. She currently works at the analysis and projections department of Statistics Austria. Her research focuses on the effects of intensified job demands on employee well-being and motivation as well as on the role of cognitive appraisal in stress processes.

Johanna Bunner is a doctoral researcher at the Faculty of Psychology, University of Vienna. Her current research focuses on safety engineers as a key factor in improving occupational safety. Special attention is dedicated to the complexities of interpersonal collaboration between managers and safety engineers (i.e. efficiency of safety engineers' consulting strategies). She is researching safety engineers' impact on safety climate and objective accident numbers.

Chapter 5
Bridges Over Breaches: Agency and Meaning as Resources for Dealing with New Career Demands

Irina Nalis

5.1 Introduction

The nature of careers has changed: today's careers are less predictable and less stable due to the fact that they last longer while allowing for decreasing lengths of individual career episodes (De Vos and Van der Heijden 2015). As described by Flecker, Fibich and Kraemer in the Chap. 2, a dynamic restructuring of work and employment is a widespread phenomenon. Especially subjectification has been shown as the most important aspect of the reorganization of work and as a direct effect of the socio- economic changes. As a result, in large of the labor market, employment has become less secure. According to many career researchers, decreases in job security and an intensification of learning demands has led to an increase in individual responsibility towards career self-management (see e.g., Forrier et al. 2015; Bimrose and Hearne 2012). Furthermore, over the course of working life, individuals face more complex and competitive careers as the traditional career contract of linear and upward moving careers has been broken and the predictability of careers has almost ceased to exist (Akkermans et al. 2016). This requires frequent changes in and outside of organizations; hence demands for the self-management of careers have risen. Given these circumstances, it is important to gain a better understanding of the resources for career self-management as those resources might help individuals build bridges over the breaches they encounter throughout their working life. When looking at resources for career self-management, I will apply what has become the classic definition of careers by Arthur and colleagues (in Dries, Pepermans & De Kerpel, 2008: 908), who defined "*career as the evolving sequence of a person's work experiences over time*".

I. Nalis (✉)
Faculty of Psychology, Universitätsstraße 7, 1010 Vienna, Austria
e-mail: irina.nalis@univie.ac.at

C. Korunka and B. Kubicek (eds.), *Job Demands in a Changing World of Work*,
DOI 10.1007/978-3-319-54678-0_5

Many of the various competing career concepts of today contain assumptions with regard to agency and meaning. Agency, on the one hand, mostly refers to self-management of a career path, e.g., via self-directed career moves inside and outside organizations [see e.g., the seminal concept of the boundaryless career]. Meaning, on the other hand, can work as an internal compass and sometimes drives a person towards the fulfillment of a calling (see e.g., the new developments in research on sustainable and meaningful careers by, De Vos and Van der Heijden (2015), Valcour (2015), or Hall and Chandler (2005) on "The career as a calling"). One career concept that combines the two dimensions of agency and meaning is the protean career (see, e.g., Herrmann et al. 2015; Baruch 2014; Gubler et al. 2014; Hall 2004). According to Hall (2004), who first introduced the concept, the protean career is managed and modified by the individual in a self-directed way and oriented toward personal values and meaning. This chapter will present an exploration of how agency and meaning offer psychological resources in dealing with the new demand of career self-management. The protean career will serve as a focal point for this endeavor yet will be accompanied by other resources that might be helpful in dealing with new career demands.

The setup of the chapter is as follows: first, breaches in today's career contracts will be examined. As part of this, the state of research on new career concepts in general (Baruch et al. 2015; Gunz et al. 2011) and on recent developments in career research with regard to sustainable careers (Van der Heijden and De Vos 2015) will be introduced. In this first section, reasons for the rise of new career concepts will be discussed and light will be shed on the dark sides of new careers. In the second section, our current understanding of the protean career concept (e.g., Herrmann et al. 2015; Baruch 2014; Gubler et al. 2014; Grote and Hall 2013) will be presented and deepened by a look at its potential to deal with new demands in working life. The focus herein will lie on the exploration of the potential of the protean career to serve as a bridge to overcome breaches in one's career path. The third section deals with ways how individuals can create personal career paths, i.e. how they can build bridges over career breaches. Within this section, further resources will be presented such as career resilience (Lyons et al. 2015), and the new concept of career crafting (Akkermans and Tims 2016; Valcour 2015). Finally, the role of career change for self-directed and values-oriented career management will be discussed.

The theoretical overview in this chapter will be deepened by qualitative interview data obtained from mid-career individuals. The interviews were conducted as part of a research project focusing on agency and meaning as resources for coping with the demands of career self-management. The interview topics were based on the framework of the protean career. The aim of interpreting and displaying the career narratives is to present possible ways of dealing with the new demands of career self-management. Hence, this subsection aims to offer insights into how mid-career professionals actually build bridges over career breaches.

5.2 The Breach of the Career Contract

5.2.1 From Organizational to Individual Responsibility

New careers differ vastly from what earlier careers and their definition of success looked like. Today, careers are increasingly characterized as personal paths with peaks and valleys. Individual career development is viewed as multifaceted, unstable, cyclical, and transitional over the life course (Bimrose and Hearne 2012). In reference to the chapter by Flecker, Fibich and Kraemer changes to work organization and the erosion of previous boundaries often lead to expanded flexibility and freedom, in particular for highly skilled workers. This tendency is often referred to as the subjectification of work. They further point out to what is in psychology often referred to as the broken psychological contract: the increasing shareholder-value orientation and the focus on short-term profits have severely affected employment relations and management control in what was termed 'disconnected capitalism' (in Flecker, Fibich and Kraemer, Chap. 2). Also macro-economic trends such as the "financialization" in the Western world, globalization and the development of international networks with their far-reaching effects, like the redistribution of work into other countries; and informatization and digitalization do have a strong impact on the development of careers both on an individual as well as the organizational level. The most notable difference between new and traditional careers can be found with regard to perceptions and definitions of career progress and success (Baruch and Vardi 2015), especially as career success was often used as the main criteria to describe careers. According to Baruch and Vardi (2015) traditional career progress followed a structured path, yet this path often plateaued: not everyone could climb the corporate ladder to the top. Moreover, this traditional career model mainly applied to Western, white men, while women and minorities were often excluded or hit glass ceilings (Baruch and Vardi, 2015).

Nowadays, we are witnessing an increase in self-directed career management where individuals take on responsibility for their career advancement. The development of careers today is increasingly becoming a task for the individual rather than an organizational task. The causes of this shift have been the subject of intense debate among career scholars. One reason might be that organizations are providing fewer opportunities for upward mobility, often as a result of an increase in market pressures in a globalized economy. Another aspect which is identified as a driver of this development is a breach in the psychological contract. The psychological contract describes the relation between employees and organizations and influences individual and organizational performance in many ways (Ng et al. 2013). It shapes the set of beliefs individuals carry about their employer and their expectations of advancement within the organization.

Hall (1996) claims that the breach in the career contract between individuals and organizations is a main driver of the shift of career management from organizations to employees. According to Sullivan and Baruch (2009), employees realize that the former ideal of a career that moves upward in a linear manner towards success is no longer attainable. They also observed changes in employees' views on the employer-employee relation, with employees increasingly viewing the relation as a transactional rather than an emotional one: formerly existing bonds were no longer present; the paycheck served as the central bond between employer and employee, with other bonds of loyalty largely absent. One reason for this development might be the decrease in organizational engagement with regard to helping individuals achieve long-term career goals by training and retaining them, as organizations increasingly practice short-term "hire and fire" policies. Hence, the breach of the career contract might lead to negative outcomes for both sides, as according to Ng et al. (2013), as is linked to various negative individual and organizational outcomes (e.g., the loss of social resources, decrease in exchange among co-workers).

However, the breach in the psychological career contract and decreasing opportunities for upward career mobility might not be the only reasons why individuals have taken on the responsibility for their careers themselves. As Ibarra (2003) argues, today we can also observe an increase in individual reflection about work and life goals. In this vein, Sullivan and Baruch (2009: 1543) argue in their review of career research that "*individuals are driven more by their own desires than by organizational career management practices*". Moreover, today's employees tend to identify with their profession more than with their organization (Baruch and Vardi 2015). This might also result in less stable relationships between organizations and employees and shorter retention spans within organizations (Akkermans et al. 2016).

Sargent and Domberger (2007) extend the view of the breach in the career contract from an organizational to a societal perspective, observing a change in younger generations' expectations of career outcomes. In this case, the perception of a breach in the psychological career contract has an influence even on individuals who have not yet experienced a breach from a past employer but rather perceive it in a more general way. According to Sargent and Domberger (2007), younger generations display a more pessimistic view with regard to their prospective material success in comparison with previous generations. Furthermore, Sargent and Domberger (2007) describe an effect of this pessimistic view on individual career proactivity, meaning that the road to individual agency might be blocked. This is of particular concern, as one prevalent assumption with regard to career management is that individuals are, or should be, increasingly proactive in their careers in response to wider economic, societal, and technological developments (see e.g., Gubler et al. 2014). To cope with these changes, the individual needs to be a continuous learner and feel empowered as an agent of her/his own career (Hall 2004; Sargent and Domberger 2007). Furthermore, the breach in the career contract demands the development of a strong sense of self that can serve as an internal compass and helps ensure that a person does not get lost on the winding road that characterizes today's working life. This leads to the question of how individuals can translate their desires into effective career management strategies. As Baruch and

Vardi (2015) summarize in their critical examination of contemporary careers, the responsibility for steering a career has shifted in many ways from the organization to the individual employee. Hence, an exploration of the resources individuals can tap into when faced with the increasing demands of developing their careers over time is required.

5.2.2 The Bright and Dark Sides of New Career Concepts

New career concepts try to offer perspectives and tactics on how to deal with macro-economic changes around us by offering a view on careers that goes beyond a linear description of upward momentum. Careers seem to offer an inventory that provides individuals several directions within their career trajectories: for instance, sideways (e.g., re-training in a new occupation or expert careers without leadership ambitions) and downward movements (e.g., job loss but also self-solicited sabbaticals) are available options. Moreover, many new career concepts, as for example the concept of the protean career, seem to be transferrable to other cultures and contexts (Shen et al. 2014). According to Hall (2004), the protean career is managed and modified by the individual in a self-directed way and oriented toward personal values and meaning.

However, new career concepts are also often one-sided as there is a tendency to neglect structural conditions. As Sullivan and Baruch (2009: 1543) highlight, "*careers do not occur in a vacuum*" as individual careers are influenced by many contextual factors such as culture, economy, and the political environment. In their discussion of career studies and their lack of a thorough investigation of broader contexts, Gunz et al. (2011) call for an understanding of careers as a social and political phenomenon. Moreover, it needs to be kept it mind that the opportunities provided by new careers go hand in hand with new demands for the individual, especially with regard to autonomy and flexibility demands. According to Baruch et al. (2015), although individuals might adapt to new career demands, they still look for security and stability. Nonetheless, new career concepts often come with a decrease in stability and heightened learning demands. Therefore, several researchers (e.g., Sullivan and Baruch 2009; Volmer and Spurk 2011; Gubler et al. 2014; Hall 1996) have raised concerns that the management of one's career path is not a welcome choice for all employees.

Furthermore, Baruch and Vardi (2015) highlight in their critical review of the dark sides of contemporary careers that the positive outcomes of new career concepts are frequently overestimated. Therefore, the blessings of freedom and growth need to be regarded with caution as mentioned also in Chap. 2 by Flecker, Fibich and Kraemer, when they talk about a "mixed blessing". Accordingly, the outcomes of today's career systems are more negative than one would expect (Baruch and Vardi 2015; Rodrigues et al. 2015; Sullivan and Baruch 2009). Discrimination or abusive supervision may be such negative outcomes (Sullivan and Baruch 2009). Moreover, Baruch and Vardi (2015) mention the risk of diminished efforts on the part of the employing organization to invest in employees, which eventually influences the long-term employability of individuals.

Another frequently-voiced critique of new careers refers to their over-emphasis on individual agency. Many new career concepts place a strong emphasis on individual freedom and growth as core values (Hall 2004). Such agentic views on career development might be the result of neo-liberal ideology [Ganesh and Inkson (2010) cited in Gubler et al. (2014)]. Furthermore, Bimrose and Hearne (2012) assume that self-negotiation of one's own career path is part of the cost of living in a "risk society".

However, many of the changes between traditional and new careers are for now mostly seen on the individual level. Sullivan and Baruch (2009) highlight the discrepancy between organizational practice and individual response. They summarize in their review on new careers that positive strategies to deal with changing work environments are still lacking in most organizations: as individuals seem to be taking on more responsibility for their personal career development, they are adapting to new career concepts. On the other hand, an adequate response from organizations (e.g., the re-organization of recruitment or career schemes) is absent. Therefore, wider implications for career development at the onset of the 21st century need to be examined and applied at the individual, organizational and policy levels. In this vein, Van der Heijden and De Vos (2015) bring a fresh perspective to contemporary careers. They isolate four central elements in their definition of contemporary careers: time, social space, agency and meaning. These elements are key dimensions of the new research direction on sustainable careers. Sustainable careers are defined as "*the sequence of an individual's different career experiences, reflected over a variety of patterns of continuity over time, crossing spaces, and characterized by individual agency, here providing meaning to the individual*". Because of its multi-dimensionality and its theoretical as well as practical applicability, Van der Heijden and De Vos' definition of sustainable careers shall set the framework for the rest of this chapter. Within this chapter, it is assumed that an understanding of careers warrants a holistic perspective on sustainable careers. With this holistic perspective in mind and acknowledging the resulting limitations, the focus of this chapter lies on the agency dimension. This focus was chosen as the agency dimension on the one hand offers practical insights for individuals on how to develop their individual career paths. On the other hand, the agency dimension provides meaning in working life. Hence, it is assumed that the agency of constructing one's personal career is itself a resource for building bridges over career paths with breaches.

This subchapter on the breach of the career contract and its implications for organizations, individuals and for a career concept perspective tried to acknowledge the contextual factors and possible strains new career demands place on the individual. Therefore, the focus of the following pages lies on the resources individuals can use to shape their personal work biographies. In the next section, the question of how to build bridges over career breaches will be deepened with help from current research on the protean career as a resource with regard to career resilience, career changes and the crafting of personal and sustainable career paths. The aim is to gain insights on individual, relational and contextual resources that help individuals to deal with new career demands.

5.3 The Bridges in Career Paths: The Role of Agency and Meaning in Overcoming Career Obstacles in Different Career Concepts

5.3.1 The Potential of the Protean Career as a Resource for a Sustainable Career

The protean career is often discussed as a contemporary approach to careers and is of key interest with regard to new careers and the question on how individuals can build bridges over the various breaches they will encounter during their working life. The protean career offers a view of careers that is neither linear nor oriented toward external goals, but rather defines career as a personal path with peaks and valleys (Hall 1996). The name of the concept is derived from Greek mythology and refers to Proteus, the god who could change his shape at will. The protean career was first introduced by Hall in 1976 and since then has been further developed by many scholars and practitioners. For instance, Herrmann et al. (2015) claim that there is renewed interest in the potential of the protean career in reaction to changes in the work environment. According to Gubler et al. (2014), who provided a literature review and refinement of the concept, the protean career concept is a theoretical framework with two subsidiary elements: one is the protean career orientation, the attitudinal aspect, and the other the protean career path, the behavioral outcome. The central dimensions of the protean career are self-directedness in the management of one's career path and a value orientation toward personal meaning (Hall 2004). Furthermore, there are two other components called "meta-skills" that describe "being protean": adaptability and identity. These meta-skills should enable the individual to navigate their career through changes while maintaining a clear sense of self. Moreover, these meta-skills are supposed to help individuals deal with the frequent changes that are encountered over various "career mini-cycles". Career mini-cycles occur in time spans of approximately five years and consist of stages of exploration, trial, establishment, mastery and disengagement.

Alongside structural changes in the working world, new questions about the purpose of work are being raised: well-being seems to be gaining greater value than wealth. This also resonates with Hall's definition of the protean career, in which he claims that the goal is subjective success (e.g., well-being) rather than objective career achievements (e.g., wealth). Hall proposes, "*the path to the top is being replaced by a path with a heart*" [Hall (1996) cited in Gubler et al. (2014: 35), Hall (2004), Sargent and Domberger (2007)]. Thus, the protean career concept offers an individual perspective on careers and success that can lead to a career with a calling (Hall and Chandler 2005), which is beneficial for individuals and society. Accordingly, the protean career is linked to the individual perspective yet tries to take the relational environment (from professional to private surroundings) into account. The relational approach is shown on the one hand in the notion of subjective success, where for instance the balance of work and family domains might serve as a career goal (DiRenzo et al. 2015). On the other hand, the concept takes into account the role of significant others in career planning activities (Grote and Hall 2013). Hence, the protean career might be a strategy aimed at strengthening individual possibilities in finding and pursuing a fulfilling work life that entails and is enriched by relational needs. Furthermore, the protean career is described as a reaction to changing career pathways as it provides individuals with greater freedom and self determination (e.g., Volmer and Spurk 2011). In support, there is evidence that a protean career orientation relates positively to various outcomes at the job level as well as with regard to career perspectives: it is positively related to job and career satisfaction (Baruch 2014; De Vos and Soens 2008), work engagement, and career planning. Moreover, protean attitudes and behavior seem to help individuals deal with unemployment (McArdle et al. 2007). Hence, it can be assumed that the protean career concept has potential to serve as a resource for individuals dealing with self-management of career development. In the following section, the protean career concept will be introduced as a possible bridge to overcome the above mentioned breaches individuals face in their working lives nowadays. Emphasis will be placed on the conceptual distinctiveness and measurement of the protean career concept.

5.3.2 The Protean Career and Its Conceptual Distinctiveness and Measurement

The protean career is widely acknowledged as a central and helpful concept with regard to new careers. Nevertheless, it is sometimes questioned whether it is a distinct construct from other career-related constructs and whether this distinctiveness can be encapsulated in concise measurements. The first question often appears with regard to the construct of proactivity: Is being protean different from being proactive with regard to one's career? With regard to this, a recent investigation by Herrmann et al. (2015) showed that a protean career orientation is a predictor of proactive career behavior even after controlling for a general proactive disposition. Herrmann et al. (2015) also showed that a protean career orientation predicts career satisfaction and proactive career behaviors beyond the disposition to proactivity. Secondly, a protean career orientation can predict positive engagement in proactive career behaviors. Moreover, the protean career orientation can be considered a distinct motivational factor (Herrmann et al. 2015) that helps us to better understand the relations between personality dispositions and career outcomes. These very recent results are likewise of interest due to the fact that Herrmann et al. (2015) conducted their study among university students as well as employees.

Further questions concern differences and similarities to the boundaryless career concept. Many scholars argue that the protean career has overlaps with the boundaryless career concept in that they are both seen as new career concepts that differ from traditional careers (e.g., Gubler et al. 2014; Briscoe et al. 2006). However, the main difference between the two concepts is that the boundaryless career concept is focused on mobility. This does not exclude individuals who are "protean" from being mobile as they might show physical mobility in their career paths. Yet protean individuals could just as well pursue their protean careers within the same organization (Briscoe et al. 2006). The focus of the protean career is therefore on individual perceptions of self-directed career development, and to a lesser extent on the question of mobility. To allow for deeper understanding of the similarities and differences between these two career concepts, Briscoe and Hall (2006) examined the interplay between the protean career and the boundaryless career. They developed eight combined typologies of the protean career orientation and the concept of the boundaryless career. These combined typologies were based on the two main dimensions of the protean career concept, self-directedness and values orientation, as well as on physical and psychological mobility as described in the boundaryless career (Briscoe and Hall 2006). The results show, for example, that being values-oriented but not self-directed is a risk for individual career management and might result in a feeling of being "lost or trapped" in one's career (Briscoe and Hall 2006). In contrast, a "protean career architect" is both self-directed and values-oriented in her or his career pattern and exhibited both physical and psychological mobility.

Although the distinctiveness of the protean career concept seems to be valid, there is still a debate on whether its measurement is concise and efficient (Gubler et al. 2014). In the past, the most frequently applied measurement of the protean career was a two-dimensional 14-item scale developed by Briscoe et al. (2006), which measured self-directedness and values orientation. However, this scale was frequently criticized [especially by Gubler et al. (2014)] for its lack of theoretical underpinnings and often inconsistent use in research. In 2014, Baruch introduced a newly developed 7-item scale for the protean career. In contrast to Briscoe et al. (2006) bi-dimensional approach, this new scale operationalizes the concept as uni-dimensional. This approach and the brevity of the scale bear several advantages, for instance, with regard to its practicality for survey research, which can be very time sensitive. Most importantly, however, Herrmann et al. (2015) validated the new scale by comparing it with the old one, finding satisfying results in terms of concurrent and discriminant validity for the new shorter scale. They also applied the scale in a cultural context different from the Anglo-American one, showing its cultural transferability. These developments in research methods with regard to the protean career not only show the distinctiveness of the concept but also its actual relevance within career research and also in terms of practical implications. The agency implied in the protean career concept together with its view on subjective success serve as a theoretically strong foundation for recent observations in research on the crafting of sustainable careers.

5.3.3 Agency, the Protean Career and Career Resilience

In the current world of work, where demands are rising and the availability of stable jobs is declining, it is of utmost importance that individuals are equipped with strategies to build bridges over the various breaches they might face in their careers. One important aspect of overcoming obstacles involves strengthening individuals' employability. Therefore, it is assumed that programs designed to build career resilience and raise employability are needed in the modern career environment. In this subchapter, the protean career is assumed to be one such possible bridge. Hence, the chapter proceeds from the optimistic notion of the protean career concept as a transformational idea (Hall 2004, 1996) that provides valuable resources for dealing with career development in changing times. In their comprehensive review on resilience in the modern career, Lyons et al. (2015) examined the protean career as a "protective factor" individuals can use to enhance their career resilience. Career resilience was originally defined as "*a person's resistance to career disruption in a less than optimal environment*" [London (1983): 621 cited in Lyons et al.(2015)]. In their examination of the relation between the protean career and career resilience, found career resilience to be an outcome of the protean career. Hence, protean career behavior and attitudes might serve as protective factors for the individual. On the other hand, their analysis also revealed differences with regard to the positive impact of the two dimensions of the protean career, self-directedness and values orientation, as each dimension worked differently as a protective factor. Lyons et al. (2015) report in their study that individuals with strong values are less resilient when faced with adversity in their careers. They assume that facing adversity can cause an individual to experience dissonance in terms of her or his values, which might in turn cause them to question the efficacy of their personal values as guiding principles. Their observations correspond with the results of Briscoe and Hall (2006), who defined typologies of individuals who demonstrate self-directedness and/or a values orientation in their career path (e.g., a strong values orientation but low self-directedness would result in a feeling of being "lost or trapped" in one's career, Briscoe and Hall 2006).

Although it was shown that being strong on values orientation does not help individuals overcome situations that oppose their belief systems, self-directed behavior – the other dimension of the protean career construct – was positively associated with career resilience. Lyons et al. (2015) attribute this result to the observation that individuals who take ownership of their careers are more likely to perceive various options at their disposal because this perception enhances their employment prospects. Furthermore, Bimrose and Hearne (2012) showed that developing and enacting targeted strategies renders a person more resilient. Thus, the protean career seems useful in dealing with new demands as it helps increase individuals' resilience. However, the question remains as to how to empower individuals with a clearly defined value set to engage in career construction activities that help them find the meaning they long for in their career.

5.3.4 Crafting a Meaningful Career

Individual agency is supposed to be crucial to attaining career success. Yet as Akkermans and Tims (2016) note, it is not clear what kind of behavioral responses individuals need in order to shape their careers successfully. One possible resource for individuals to actively shape their careers according to their needs and values might be found in the job crafting literature. Recent developments in organizational studies research focus on the potential of job crafting as a means of altering jobs according to employees' preferences, sources of motivation and passions (Tims et al. 2012). Job crafting describes the shaping of daily work experiences with regard to (1) increasing job resources, (2) increasing challenging job demands, and (3) decreasing hindering job demands (Tims et al. 2012). Furthermore, Berg et al. (2013) describe job crafting as an active process in which employees redefine and reimage their job design in ways that are personally meaningful.

Akkermans and Tims (2016) define career crafting as a "*proactive behavior that individuals perform to self-manage their career, which is aimed at attaining optimal career-related outcomes*" (Akkermans and Tims 2016, p. 12). In the presentation of their newly-developed Career Crafting Survey, Akkermans and Tims draw on research on job crafting by extending daily job crafting activities into the long process that eventually shapes our careers. The recent work of Akkermans and Tims is rooted in the framework of sustainable careers as introduced above and is also designed to allow for insights into the agency dimension of sustainable careers (cf., Valcour 2015). According to Akkermans and Tims (2016), in order to study career crafting it is necessary to take not only job crafting experiences but also career competencies into account when attempting to describe "what" individuals can do to craft their careers. Career competencies are described as the sum of knowledge, skills and abilities which are necessary for career development [Akkermans et al. cited in Akkermans and Tims (2016)]. Yet they argue that despite allowing insight into a person's knowledge and skills, career competencies cannot reveal whether the actual shaping of careers is taking place. Therefore, future career planning activities—which are assumed to be strongly influenced by earlier career experiences—are another important aspect of career crafting (Baruch and Vardi 2015).

In the case of job crafting, job crafting activities allow for positive experiences in the workplace and increase meaning and purpose (Berg et al. 2013). It can be assumed that these positive experiences provide a resource of confidence, which can translate into further crafting activities – both job crafting and career crafting. While positive experiences might serve as resource, the opposite holds true for negative experiences. Baruch and Vardi (2015) argue that negative experiences in working life tend to have a far stronger impact on future career planning behavior as it is regularly described. This is of special concern with regard to future career planning activities and proactive career planning because proactive career planning has been shown to increase employability, career satisfaction and subjective career success (Forrier et al. 2015; Lyons et al. 2015; Bimrose and Hearne 2012). To increase workers' ability to deal with the new demands of career self-management, it might be necessary to enhance their opportunities to actually craft their careers. Akkermans and Tims (2016) show that individuals who actively engage in job crafting might eventually translate this ability from daily job activities to a pro-active career crafting behavior.

Having chosen agency and meaning as the focal point of my search for resources to help workers to cope with the increasing demands of today's work world, I assume that career crafting is itself a resource into which individuals can tap to build a bridge over their career breaches, as described above. Therefore, career crafting seems to follow in the tradition of career concepts such as the protean career that offer broader views on career success than the often claimed yet seldom reached linear path to the top. Thus, they provide a better understanding of what to do in order to build a path with a heart (Hall 2004).

5.3.5 Career Agent of One's Own Life: Employability Through Meaningful Career Narratives

Another possible resource for individuals to deal with the new demands of career self-management might be found in career narratives. As today's careers no longer follow the 20-40-20 pattern of 20 years of childhood and education, 40 years of – rather – stable employment, and another 20 years of retirement, it is necessary to adopt a new perspective on careers that is today often under- if not entirely un-taught in schools. With regard to this structural change, employability is of special concern. It is assumed that there is a danger of falling behind if one doesn't engage in up-skilling or re-skilling every five to ten years (Brown et al. 2012). Furthermore, it is argued that a lack of re-skilling activities might render an individual more vulnerable with regard to their personal career development perspective. In contrast, individuals who experience the transferability of their skills to new jobs enjoy advantages in terms of career advancement (Bimrose and Hearne 2012). Accordingly, many researchers point out that career adjustment requires purposeful action and ownership over one's actions (see e.g., Ibarra 2003).

A promising path towards ownership, transferability and action might be found in Ibarra and Barbulescu's (2010) investigation of major career transitions. Ibarra and Barbulescu (2010) examined the importance of career narratives in terms of allowing individuals to take on the role of active career agents. They showed that career narratives where individuals describe themselves help them link former and new aspects of their professional role. Moreover, career narratives allow the individual to experiment with the actual demands and outcomes of a desired new profession. Therefore, it is assumed that career narratives serve as yet another bridge to overcome career breaches by allowing the individual to actively deal with unexpected changes. Furthermore, it is assumed that career narratives can help individuals highlight their individual agency and provide meaning by allowing workers to use reflection and construction to shape their next career sequence. They envision and then enact the change they want to see. Moreover, Ibarra and Petriglieri (2010) examined identity play as a resource for dealing with transitions, as it allows people to experiment with possible future selves in times of transition. They showed that identity play provides opportunities for unexpected discoveries of skills, interests, and meaning, which are helpful in shaping the next sequence of a person's career path.

The aim of this section of the chapter was to present current literature and new ideas on opportunities for workers to shape their career paths by building bridges over career breaches. Furthermore, the subsection aimed to offer insights on possible strategies individuals might need to engage in career self-management. By providing an overview on current literature on the protean career concept, career resilience, career crafting and career narratives, I sought to allow for a broader understanding of recent discussions and practical implications. Nevertheless, it needs to be noted that many other aspects of career research concepts could have been discussed in this overview; in particular, Savickas and Porfeli's (2012) concept of career adaptability deserves mention, as do career self-efficacy (Betz 2007) and the theory of life design and the constructing of careers (Savickas 2012). Yet in order to focus on concepts that illustrate different resources for individuals to shape their careers, this chapter does not claim to have provided a complete overview on the vast – and sometimes cluttered (Baruch et al. 2015) – field of career concepts. Instead, the aim was to provide an idea of how the concepts of career resilience, the newly-developed concept of career crafting and career narratives might serve individuals dealing with new demands in their working lives. Therefore, this section aimed to provide greater understanding of how to foster individual strategies that allow for agency and meaning to cope with the increasing demand of career self-management. How individuals actually approach new demands in working life will be the focus of the next section.

5.4 Three Career Narratives: Peaks and Valleys Instead of "Up or Out"

This section presents aspects of career narratives of three Austrian professionals which were collected as part of a larger study on the development of the protean career path at the University of Vienna in 2015 and 2016. The sample was recruited from various working backgrounds (non-profit sector, for-profit sector). Most respondents were highly skilled, but low skilled workers were also interviewed. The aim of this subsection is to offer insight into actual protean career paths. According to the theoretical deliberations described above, the focus will lie on if and how being protean helped individuals deal with peaks and valleys in their working lives and shape their personal careers. Moreover, under-researched influences of chance events and reference groups on the development of a career path will be explored (see e.g., Gubler et al. 2014 for their recommendation that researchers explore the role of chance events and reference groups with regard to the protean career).

The three career narratives provide insight on how the interviewed individuals proved capable of transforming their professional lives over time. Each of the presented narratives shares the commonality of tremendous effort in developing a personal work identity (for the concept of work identity, see Ibarra 2003). Moreover, in all of these stories people experienced breaches yet found ways to build paths that better suited their values and were more closely tied to subjective success than what their original professions and paths would have allowed. The three career narratives were chosen to illustrate the potential of agency and meaning as resources for attending to demands for career management.

This section aims to allow for an in-depth understanding of questions of individual, relational and contextual resources that help individuals build bridges over career breaches. It starts with first experiences in the working world, gives insights on self-directed and values-oriented career moves, provides accounts of chance events and the role of reference groups, and ends with reflections on career success in particular and commonly-held beliefs about careers today in general. Thereby, the following parts of this chapter will put individuals who actively transformed their careers by changing jobs, moving to different countries, re-training for an entirely new profession or making the jump from employee to entrepreneur at center stage. Whenever a change occurs, it is of interest how the original career choice was ruled out and what drove individuals to pursue a new career path. Therefore, the career narratives presented here contain stories of career change but also focus on other aspects of how to deal with changes in the world of work.

At the beginning of each interview, participants were invited to talk about their working life in general. This approach (see also narrative interviews, Schütze 1982), allows participants themselves to choose the starting point of their story and thus of their professional life. This approach often resulted in accounts of very early, first experiences of summer jobs. These stories often revealed emerging initial interests, as they allowed participants to find out what might suit them at a very young age. Moreover, participants who started their narrative about their professional lives in their early teenage years reported that they have gained through them valuable insights about themselves and their future career paths.

5.4.1 First Steps in the Working World

This is how Anna, 35, today a journalist, described her first steps in the world of work:

> I started working very young… the first time at the age of 14 in the shop of a friend of my mothers… which was actually absurd as at that age I wasn't even allowed to work yet… but I wanted to be independent… I always enjoyed working and I was really interested in finding out what work is like … it wasn't my parents who told me to work… I had this wish myself.

Anna was also very decided on her choice of studies and recalled that her parents supported her fully in her personal choice. After taking her A-levels, she remembers having known especially about the sectors and occupations she didn't want to go into, but she also knew about her interest in journalism, which was fueled as she recalls by "*movies and TV shows from the 80s and 90s*". She further reported:

> Then I came to Vienna and I knew the only thing that really interests me is journalism and the only thing close to this in Austria was to study communications in Vienna… and while queuing for my application I realized that I needed to inscribe for a second subject… and so I decided on politics out of a great interest… however, the studies were not what I expected… no practical implications at all and 13 times a day we were told that if you want to become a journalist you don't belong here… then I started to pick classes that were closer to journalism and there I encountered guest professors and lecturers from Germany who were really good and taught me for instance about war journalism.

It was also about this time when she told herself:

> I need to stop working as a waitress, I need to gain relevant work experience and so I looked up the university's job portal where I found an announcement …for an assessment center for a junior enterprise…I went there and I really enjoyed the atmosphere…everything was self-organized…I had never appreciated those strict hierarchies where the youngsters are also treated as the dumber ones ….at this enterprise everyone could make up their own rules …this is where I learned for the first time to structure my work and to look at an organization as a whole…and to understand more the interplay of an entire team.

After her studies, she could not break into the field of journalism, so she accepted an offer from a consulting firm she knew from a cooperation during her time at the junior enterprise.

> It didn't interest me at all but I found my team leader very competent and I had the feeling from the beginning this is someone I can learn a lot from… but he left after a year … later I joined another team… where we would organize our trainings … I really enjoyed that time

> because it was again a thing of team spirit … but then in an assignment abroad I started to think again… I actually still want to become a journalist… this has always been my dearest wish from the bottom of my heart.

Hence, a job that provided Anna stability yet lacked the meaning she sought in her working life did not fulfill her needs and left her uncertain about her current career path.

Other participants in the qualitative study focused the starting point of their story on their decision-making process in terms of education. Here, the role of family members was often more dominant than participants would have wished. For example, Sarah, 35, advertising professional, described this process of choosing her field of studies:

> To me communication was always interesting, my father was a media artist … creative work was always interesting. My father said, "don't study design and don't aspire to become a designer" … and well… at the age of 18 no one is certain what to do, right?… and then everybody was happy when I announced that I am going to study business… once there I specialized in marketing and advertising thinking that through this side entrance I will get to the creative side… and somehow this made all of us happy.

Or Stefan, 45, entrepreneur today, originally trained as an electro-mechanic[1]:

> I come from the countryside and I did exactly what my colleagues from primary school did… I decided to do vocational training instead of going to a college-prep high school… I did actually look a bit around for a creative job but as many of my family members were electricians… my father, my uncle, my cousin … it was presented to me that this would be a good choice as it provides a stable income … to nourish a family and so on … so I pursued their path… but I never developed a real passion for it.

As these accounts show, neither Sarah nor Stefan were self-directed in their first choice of profession. Their ability to be agentic and proceed self-directedly through further steps in their career paths developed through chance events. As the following parts of their narratives will show, Stefan's path was very influenced by chance events. In contrast, Sarah's account of a disappointment in her first job in the communications industry that accelerated her first major self-directed career move provides an example of agency as way to overcome career obstacles:

> My first boss was the first leader I actually got to know … and this was a woman who as completely overwhelmed by her job… she drank alcohol during the work day … and I was just overwhelmed, too. I was really done with everything… and I started to think of ways to get out of it. Where am I going to learn something? Because nobody here is showing me anything … and so off I went to America.

---

[1]Note on the contextual factor of the Austrian educational system: a dual system exists in Austria in which students can complete an apprenticeship for jobs e.g., in sales or skilled trades that combines work in the private sector and public schooling once a week.

5.4.2 Self-directed and Values-Oriented Career Moves Triggered by Chance Events

Stefan's entrance into the world of work was, as was shown earlier, influenced by family and peer expectations. Yet he soon discovered that he was not living according to his own choice. Due to an unpredicted and unplanned chance event – an accident outside of work – he was allowed to re-train for another profession, thus escaping his dreaded job:

> During my second year it became clear to me that I don't want to continue with this job… but this is how things go there … what you start you'll finish but straight after I decided to finish my apprenticeship I injured my knee severely and this gave me the opportunity to train for another profession. This time it was to become an office clerk… during that re-training I was also actively involved in voluntary work with children and teenagers … I also started to consider looking for a job in social work with children and youth.

However, this was not the only time Stefan's professional life was crucially influenced by an unplanned event which he eventually translated into self-directed action. Stefan's second chance event allowed for a big step up the hierarchical ladder and provided lots of "freedom and growth" [core values of the protean career according to Hall (2004)], and hence for values-oriented and self-directed career advancement. Furthermore, as Brown et al. (2012) also showed in their qualitative exploration of mid-career career changers, this sample also exhibited positive attitudes towards learning through challenging work:

> After I started to work at a child care organization the managing director became very ill and had to go on leave … I was only 21 at that time and was asked since there was no one else around… if I could imagine taking over the responsibility of managing director… well this is how I became the managing director of this organization for ten years… it was a very dynamic time. I was young and brought many other young people into the organization. And when it came to training methods I could always swap views with friends of mine who worked in the private sector and shared a lot of knowledge and methods with me. For instance, I could define the standards for qualifications for heads of departments, and we designed a whole new management program for human resource development… This organization grew from let's say 13, 14 to over 250 people during those ten years.

However, this considerable success in terms of objective measures of responsibility did not satisfy Stefan over the long run. His desire for a change arose again:

> After about 8, 9 years I became a bit bored …and I started to consider quitting entirely and getting into organization consulting …the development of organizations was my hobby so to say …but at the time …actually by accident again it became public that the managing director of the national organization in Vienna was about to leave and he asked me if I would take over his job as I was anyways prone to change.

In line with Sargent and Domberger's (2007) qualitative exploration on the development of a protean career orientation among undergraduates, a violation of values often jumpstarts reflection and subsequently action to change something in one's career. However, this notion only holds true for individuals who are driven by agency and meaning (e.g., Volmer and Spurk 2011; Briscoe and Hall 2006 showed that being values-oriented but not self-directed can even inhibit career proactivity and the shaping of a protean career path). An example of a contrast between the beliefs one holds towards one's own profession and towards one's employer, and therefore an example of values violation, was described by Sarah:

> It was very sobering to realize that at the core of the business it is not about the client or the thing itself …but that it is about some …interests of a shareholder network somewhere in North America which I have nothing to do with … I am no one special, it is not about my development, it is only about shares and personal enrichment. This was so sobering … so sobering to see that the work of world is not about creating something together but only about self- interest and other dimensions which have actually nothing to do with the main business.

Anna also had to undergo a rupture in her career path, which eventually turned out to be the chance event she needed to get back on track of pursuing her original calling. At that time, she had already been working for four years in business consulting but felt the urge to get her foot into journalism. However, she received no response to her applications in the media world. At the same time, she was rewarded by her consulting firm employer as one of its best performing team members:

> There you get grades like in school… it really is ridiculous… but my degree was excellent… with a big ceremony on stage and all.

At the same time, there were some intense discussions about a 2-month holiday that had already been agreed upon and that her supervisor had signed off on. She nevertheless went on the trip and when she came back to work, she learned that she had been fired, which eventually also keyed up her agency to find the meaningful job she had always dreamed of:

> It was the best thing that could have happened… I wouldn't have had the guts to resign from my job without having a new assignment yet working so many hours at that consulting firm didn't give me enough time to conduct a job search and write applications for work as a journalist.

Stefan, despite working in the social sector, also reported experiences that contradicted his values and his drive to structurally change the organization for what he believed to be for the better. Hence, he experienced a decrease in meaning and was hindered from performing as an active agent in the organization. In retrospect, he reasoned:

> Now as a self-employed person I realize what disturbed me most in all these organizations … although you can actually have an impact and much more of it as for example in a position in administration it felt to me that 90% of energy is being wasted because someone obstructs the process … I mean those destructive forces.

The presented accounts give an insight into how self-directed career moves are linked to challenges with regard to both agency and meaning, especially when participants are confronted with organizational practices that are in contradiction to their values.

5.4.3 Reference Groups and Their Role in and Effects on Career Change

Grote and Hall (2013) highlighted the importance of reference groups when examining career paths. In a 1996 handbook edited by Hall, "The career is dead – long live the career", the common theme that emerges is a relational perspective on careers. However, this perspective is often paid insufficient attention in career research today (Gubler et al. 2014). Therefore, these interviews place an emphasis on reference groups.

For Stefan, the most profound change with regard to his career change was relating to his family and the need to adopt a new way of dealing with new roles and financial insecurities, which were formerly unknown in his family.

> Relatives played a very small role, a very small number were surprised but didn't know what was going on. What was much more difficult was the inner-family context. Responsibilities shifted, I transferred my place of work to my home, the house had to be renovated … it was a really active time…while my wife started to work more as our children were at that time old enough. Altogether it is very exhausting for a partner, this change from a situation of security to a state of insecurity. This was difficult for all of us and for me, too. It is something you have to get used to … these insecurities.

Anna experienced the end of a long-term relationship during her transition from consultant to journalist. As it became more clear that her endeavor to enter journalism would possibly mean she has to stay longer in Berlin, her five-year relationship with her partner deteriorated and eventually ended. This experience, she recalled, showed her for the first time the ways in which one's partner and eventually one's entire relationship are affected by career decisions:

> I was working as an intern in Berlin at that time… and my partner at that time originally had accepted the fact that I would spend a few months there…it left an imprint on me to realize that a partner can't always accept all the decisions that you make for yourself.

These accounts show that although at the beginning of one's working life relatives might play an important role, the impact of reference groups seems to lessen with greater age and experience. This was mostly shown in that during the narrative interviews, as proposed by Schütze (1982), the stimuli questions used to elicit career narratives did not explicitly ask about the role of reference groups. Nevertheless, individuals were able to choose what and whom to include in their narratives, i.e. whether they placed an emphasis on family and friends or not. In the interviews, the role of friends and spouses as active enablers or hinderers of career decisions was rarely mentioned at all initially. Therefore, the interviewer asked additional questions about the role of family, other relatives, partners and friends. Interviewees' answers showed that family did play a role, yet a lesser one than expected by the interviewer and suggested in the literature (Grote and Hall 2013). Moreover, the accounts show that friends were of very little or no relevance to the narratives. Nevertheless, it needs to be taken into account that the individuals portrayed here displayed a very high protean career orientation, and it might be assumed that such individuals have a tendency not to let others interfere with their career plans.

5.4.4 *Meaning Over Money? Exploring the Perception of Subjective Success*

An issue of special concern for working biographies in general and in studies of the protean career concept in particular is the question of success. As Gubler et al. (2014) note, career success from the point of view of subjective success (the term psychological success is also used synonymously) is the most frequently studied aspect of the concept. In its original form, the idea of subjective success is contrasted to objective career measures such as money or fame and re-focuses the idea of success on well-being and meaning. According to Hall (1996), breaches in career contracts have led to the re-evaluation of careers and career success. A recent cross-cultural study in 11 countries by Shen et al. (2014) supports this notion of a shift toward self-determined definitions of career success. Furthermore, it was shown that an increasingly large population of individuals define their success in terms of psychological achievements (Ng and Feldman 2014). Within the framework of the protean career, Hall proposes, "*the path to the top is being replaced by a path with a heart*" [Hall (1996) in Gubler et al. (2014: 35)]. Thus, the protean career can also be regarded as a resource in that it offers a different perspective for careers and allows workers to be guided by an inner compass rather than external incentives. Additionally, Hall and Chandler (2005), claim that this individual perspective on careers and success can lead to a "career as a calling", which is beneficial to individual and society.

Here is Stefan's account of his subjective goals and values:

> I wanted to have an impact in society, I wanted to change things via politics … money really never played a role in that. But I have to add that I always had enough … well I started working at age 15, so the salary I earned as an electro-mechanic while school students had to get along with their pocket money. I even felt privileged and later as a managing director I didn't earn so badly. So well, all my professional life at that time was about idealism and the drive to change something but more and more the perspective of maybe turning a tiny wheel or having an impact in policy making didn't sit well with me. I felt more inclined to do something … so my idea of a new way of travelling, an eco-sustainable way of travelling. This is a product I can fully relate to my values.

Anna's account is in the vein of the discussed balance between objective and subjective career measures. Hall (1996) highlights that subjective career goals do not stand in opposition to organizational goals but rather enable both psychological as well as entrepreneurial success. Volmer and Spurk (2011), further support this notion by combining both subjective and objective success as criteria for overall success. In her reflection on success, Anna openly spoke about the dilemma of doing a job she loves and believes to be relevant to society but where she is unhappy with her current employer. She further spoke about the financial problems she would face were she to change employers now, which left her to date undecided on where to go next:

> Journalism is the most exciting job out there. To engage with politically and societally relevant proceedings… I actually find journalism pretty selfish because you can spend your whole time working on things that you are interested in as a private person, too.

A new perspective arose for Anna when an independent, investigative news platform she had already worked with offered her a job. However, she said that this job was very poorly paid and she now looks for opportunities to unite her path with a heart with her needs in terms of financial security and in terms of balancing her life domains (see e.g., DiRenzo et al. 2015):

> Honestly we want to live differently. I really would love to go on a sabbatical for about three to six months were I could read tons or start an educational program. I would really like to get off the hamster's wheel but where I was offered another job they really earn nothing. My life style is really not exaggerated but I don't want to live so close to the poverty line. I don't think I could do it again but maybe I could find some things I could work on in addition to the job there.

Sarah didn't actually articulate whether she would forsake financial benefits for a "path with a heart" (Hall and Chandler 2005). However, she expanded her reflections to a general observation on the quest for meaning in working life. She came up with another observation on whether the search for meaning in working life is a relevant question at all, which she whole-heartedly defended:

> And when I then hear about so called luxury problems in terms of searching for meaning in our jobs…well… isn't it the aspiration as a cognitive being to search for meaning in one's life? If not I really wonder what's it all about.

This section on career narratives ends with an example of the reflection that might be triggered in interviews on personal work experiences (Brown et al. 2012). The following remark by Sarah challenges some commonly-held beliefs in the arena of self-help and commercial career counseling.

> Get out of your comfort zone, experience new things, change yourself, be flexible… the big credo of flexibility… this is the new ethics of work but during my MBA program I thought being flexible and doing something new and getting out of your comfort zone is actually not what is good for people … I would rather say it is extremely difficult to establish a comfort zone in the first place, which is what I believe to be the real success. Because we have to question where this whole flexibility is going to take us?… Instability leads to uncertainty … and this doesn't make you happy in the long run.

This remark resonates with the position of many scholars highlighting the need to question the overly positive view of new careers (Baruch and Vardi 2015; Baruch et al. 2015). Furthermore, it is in line with Sennett's (1998) classic work on the "corrosion of character" and the dangers of excessive flexibility demands. Hence, the exploration of the protean career as a resource in dealing with new demands presented here tries to strengthen our understanding of the potential of self-directed and values-oriented career moves yet tries to stay balanced in order to not overly promote the case for change. Nevertheless, the perspective of the protean career as a potential bridge over career breaches is strengthened by the respondents' accounts above.

5.5 Conclusion

This chapter aimed to provide an overview of recent trends in the field of career studies and present fresh perspectives for dealing with new demands in the world of work. The main focus was on potential resources in dealing with change. Agency and meaning as described in various new career theories and especially in the influential protean career concept were chosen as a focus to examine how individuals can manage their career in a self-directed and values-oriented way. Moreover, the newly introduced concept of sustainable careers was chosen as a broader framework for understanding the complexity of contemporary careers (Van der Heijden and De Vos 2015; Akkermans et al. 2016; Akkermans and Tims 2016). Additionally, career crafting (Akkermans and Tims 2016), career resilience (Lyons et al. 2015; Bimrose and Hearne 2012) and career narratives (Ibarra and Barbulescu 2010; Ibarra and Petriglieri 2010; Ibarra 2003) were presented as stabilizing forces in an unstable world of work.

My goal was to offer insights on how individuals can overcome the various breaches they might encounter during their working lives and discuss what can be done to build bridges that eventually lead to personal career paths. This theoretical approach was deepened by the presentation of three career narratives. The narratives were presented with regard to dimensions of the framework of the protean career and also embedded in the focus on agency and meaning as potential resources for a subjectively successful career. To examine the potential of agency and meaning as resources in dealing with new demands in the world of work, actual career patterns and perceptions of self-determined and values-oriented career moves were investigated. Fostering or hindering conditions and the view of subjective success were also part of the three career narratives. The interview data allowed for an understanding of the experiences and choices individuals report when faced with changes in their working lives. Moreover, the presented narratives show how dimensions of the protean career concept served as a resource for the interviewees in coping with these demands. For instance, self-directed career moves seemed to support the active shaping of careers and to increase perceived employability. With regard to meaning, it was shown that clear values helped the interviewees overcome work situations where their individual needs were not met. Furthermore, chance events were mostly described as welcome catalyzers for self-initiated career changes. Additionally, reference groups were described as decreasing in importance with age in terms of career choices, yet they often provided new career opportunities.

Implications include several external resources that should be further developed in the future in order to reflect the changing nature of careers, as well as the further development of internal resources to allow individuals to increase their employability and subjective career success through agency and meaning. The following fields seem to be fruitful for further investigations as well as for HR practitioners' development of new management styles:

External resources include:

- Educators could and should play an effective role in helping students assess who they are. Furthermore, educators should help them develop reflection skills and use feedback tools to initiate enhanced reflection that is not limited to looking backwards at the end of a process but involves reflecting throughout the learning cycle.
- Organizations should develop career development schemes that better reflect non-linear moves e.g., allowing for more specialized careers rather than having experts either pushed into the role of team leader or plateau in a hierarchical position where they need to follow the instructions of lesser specialized team supervisors.
- Colleagues can take on the role of peer coaches (see e.g., Parker et al. 2015) and help detect weaknesses and strengths that might foster further career moves.
- Close bonds to friends and family members can also serve as a sounding board and provide feedback, especially with regard to one's talent and the search for purpose and meaning in working life beyond traditional career paths as they might know about one's personal desires.
- Weak ties, as described in the classic study by Granovetter (1973), will possibly become even more important for gaining knowledge about possible further job opportunities but moreover by allowing a person to gain insight into different sectors or work roles that might seem interesting.
- Supportive societal circumstances, such as the Austrian model of the "Bildungskarenz", a state-sponsored sabbatical where employees can take up to 12 months of paid leave to up- or re-skill (e.g., by obtaining degrees; starting a study program). The international implementation of this model and its extension to the self-employed would help increase employability over the entire course of today's and tomorrow's longer careers.

Internal resources include:

- Transformational career attitudes (Briscoe and Hall 2006) which are high in values orientation and self-directedness have been shown to be dynamic and especially helpful in times of unemployment.
- Trying on possible future selves (Ibarra and Petriglieri 2010) through experimentation can allow for a better understanding of one's existing talents, gaps in knowledge or personal strengths. The exploration of identity through work and play combines experience with preparation for transitions.

Limitations of the career accounts presented here include the focus on mid-career professionals and the sample criteria of change or rupture. However, as career changes are on the rise for various reasons, I believe that it is of equal scientific and practical relevance to further explore potential resources in dealing with new demands and further investigate the potentially positive (and/or negative) effects of individual career crafting.

References

Akkermans, J., De Vos, A., & Van der Heijden, B. (2016). Going the distance: Conceptualizing and examining the sustainable career, Academy of Management Submission No. 14372.

Akkermans, J., & Tims, M. (2016) Development and validation of the career crafting survey, AOM Career Division Abstract, Academy of Management Submission No. 14372.

Baruch, Y. (2014). The development and validation of a measure for protean career orientation. *The International Journal of Human Resource Management, 25*, 2702–2723.

Baruch, Y., Szücs, N., & Gunz, H. (2015). Career studies in search of theory: The rise and rise of concepts. *Career Development International, 20*, 3–20.

Baruch, Y., & Vardi, Y. (2015). A fresh look at the dark side of contemporary careers: Toward a realistic discourse. *British Journal of Management*, 1–18.

Berg, J. M., Wrzniewsky, A., & Dutton, A. (2013). Perceiving and responding to challenges in job crafting at different ranks: When proactivity requires adaptivity. *Journal of Organizational Behavior, 31*, 158–186.

Betz, N. E. (2007). Career self-efficacy: Exemplary recent research and Emerging Directions. *Journal of Career Assessment, 15*(4), 403–422.

Bimrose, J., & Hearne, L. (2012). Resilience and career adaptability: Qualitative studies of adult career counseling. *Journal of Vocational Behavior*, 338–344.

Briscoe, J. P., & Hall, D. T. (2006). The interplay of boundaryless and protean careers: Combinations and implications. *Journal of Vocational Behavior, 69*, 4–18.

Briscoe, J. P., Hall, D. T., & DeMuth, R. L. F. (2006). Protean and boundaryless careers: An empirical exploration. *Journal of Vocational Behavior, 69*, 30–47.

Brown, A., Bimrose, J., Barnes, S. A., & Hughes, D. (2012). The role of career adaptabilities for mid career changers. *Journal of Vocational Behavior, 80*, 754–761.

De Vos, A., & Soens, N. (2008). Protean attitude and career success: The mediating role of selfmanagement. *Journal of Vocational Behavior, 73*, 449–465.

De Vos, A., & Van der Heijden, B. I. J. M. (2015). *Handbook of research on sustainable careers*. Cheltenham: Edgar Elgar Publishing.

DiRenzo, M., Greenhaus, J. H., & Weer, C. H. (2015). Relationship between protean career orientation and work life balance: A resource perspective. *Journal of Organizational Behavior, 36*, 538–560.

Dries, N., Pepermans, R., & De Kerpel, E. (2008). Exploring four generations' beliefs about career Is "satisfied" the new "successful" ? http://doi.org/10.1108/02683940810904394

Forrier, A., Verbruggen, M., & De Cuyper, N. (2015). Integrating different notions of employability in a dynamic chain: The relationship between job transitions, movement capital and perceived employability. *Journal of Vocational Behavior, 89*, 56–64.

Granovetter, M. (1973). The strength of weak ties. *American Journal of Sociology, 78*, 1360–1380.

Grote, G., & Hall, D. T. (2013). Reference groups: A missing link in career studies. *Journal of Vocational Behavior, 83*, 265–279.

Gubler, M., Arnold, J., & Coombs, C. (2014). Reassessing the protean career concept: Empirical findings, conceptual components, and measurement. *Journal of Organizational Behavior, 35*, 23–40.

Gunz, H., Mayrhofer, W., & Tolbert, P. (2011). Career as a social and political phenomenon in the globalized economy. *Organization Studies, 32*, 1613–1620.

Hall, D. T. (1996). *The career is dead—Long live the career: A relational approach to careers*. San Francisco: The Jossey-Bass Inc.

Hall, D. T. (2004). The protean career: A quarter-century journey. *Journal of Vocational Behavior, 65*, 1–13.

Hall, D. T., & Chandler, D. E. (2005). Psychological success: When the career is a calling. *Journal of Organizational Behavior, 26*(2), 155–176.

Herrmann, A., Hirschi, A., & Baruch, Y. (2015). The protean career orientation as predictor of career outcomes: Evaluation of incremental validity and mediation effects. *Journal of Vocational Behavior, 88*, 205–214.

Ibarra, H. (2003). *Working identity: Unconventional strategies for reinventing your career*. Harvard: Business School Press.

Ibarra, H., & Barbulescu, R. (2010). Identity as narrative: Prevalence, effectiveness, and consequences of narrative identity work in macro work role transitions. *Academy of Management Review, 35*, 135–154.

Ibarra, H., & Petriglieri, J. (2010). Identity work and play. *Journal of Organizational Change Management, 23*, 10–25.

Lyons, S. T., Schweitzer, L., & Ng, E. S. W. (2015). Resilience in the modern career. *Career Development International, 20*, 363–383.

McArdle, S., Waters, L., Briscoe, J. P., & Hall, D. T. (2007). Employability during unemployment: Adaptability, career identity and human and social capital. *Journal of Vocational Behavior, 71*, 247–264.

Ng, T. W., & Feldman, D. C. (2014). Subjective career success: A meta-analytic review. *Journal of Vocational Behavior, 85*, 169–179.

Ng, T. W. H., Feldman, D. C., & Butts, M. M. (2013). Psychological contract breaches and employee voice behavior: The moderating effects of changes in social relationships. *European Journal of Work and Organizational Psychology, 23*, 537–553.

Parker, P., Wasserman, I., Kram, K. E., & Hall, D. T. (2015). A relational communication approach to peer coaching. *The Journal of Applied Behavioral Science, 51*, 231–252.

Rodrigues, R., Guest, D., Oliveira, T., & Alfes, K. (2015). Who benefits from independent careers? Employees, organizations, or both? *Journal of Vocational Behavior, 91*, 23–34. http://doi.org/10.1016/j.jvb.2015.09.005

Sargent, L., & Domberger, S. R. (2007). Exploration of the development of a protean career orientation values and images violations. *Career Development International, 12*, 545–564.

Savickas, M. L. (2012). The 2012 leona tyler award address: Constructing careers—Actors, Agents, and Authors. *The Counseling Psychologist, 41*(4), 648–662.

Savickas, M. L., & Porfeli, E. J. (2012). Career adapt-abilities scale: Construction, reliability, and measurement equivalence across 13 countries. *Journal of Vocational Behavior, 80*(3), 661–673.

Schütze, F. (1982). *Berufsbiographieforschung und narratives Interview*. Kassel: Gesamthochschule Kassel FB Sozialwesen.

Sennett, R. (1998). The corrosion of character: The personal consequences of work in the new capitalism, New York: W.W. Norton & Co.

Shen, Y., Demel, B., Unite, J., Briscoe, J. P., Hall, D. T., Chudzikowski, M., et al. (2014). Career success across 11 countries: Implications for international human resource management. *The International Journal of Human Resource Management*. doi:10.1080/09585192.2014.962562

Sullivan, S. E., & Baruch, Y. (2009). Advances in career theory and research: A critical review and agenda for future exploration. *Journal of Management, 35*(6), 1542–1571. http://doi.org/10.1177/0149206309350082

Tims, M., Bakker, A., & Derks, B. D. (2012). Development and validation of the job crafting scale. *Journal of Vocational Behavior, 80*, 173–186.

Valcour, M. (2015). Facilitating the crafting of sustainable careers in organizations. In A. De Vos & B. I. J. Van der Heijden, (Eds.), *Handbook of Research on Sustainable Careers,* (pp. 20–34). Cheltenhams: Elgar.

Van der Heijden, B. I. J. M., & De Vos, A. (2015). Sustainable careers: Introductory chapter. In A. De Vos & B. I. J. Van der Heijden, (Eds.), *Handbook of Research on Sustainable Careers,* (pp. 1–19). Cheltenhams: Elgar.

Volmer, J., & Spurk, D. (2011). Protean and boundaryless career attitudes: Relationships with subjective and objective career success. *Journal for Labour Market Research, 43*, 207–218.

Author Biography

Irina Nalis is a doctoral fellow at the Department of Applied Psychology at the University of Vienna and a member of the work group "Work in Times of Change." Her research is centered on career self-management. Currently, she is studying the relation between career-related demands, career crafting, career resilience, and subjective success. She is also interested in the prevalence of career self-management in the broader public and antecedents of meaningful careers.

Chapter 6
New Ways of Working and Satisfaction of Psychological Needs

Cornelia Gerdenitsch

6.1 Introduction

Financialization, network economy, and digitalization are three central socio-economic developments that have transformed work and employment for example towards an increased autonomy for workers (see Flecker et al., Chap. 2). Highly skilled workers increasingly use information and communication technologies, which enable them autonomy about when and where they do their work (Demerouti et al. 2014; Hill et al. 2008). As a result, various forms of flexible work have emerged, such as flextime (choosing the beginning and end of the work day around a set of core hours), flexplace (including teleworking, i.e. working from outside the office on a regular basis, with telecommuting or home offices representing a form of teleworking; Sullivan 2003), or nomadic work (which involves a total lack of temporal and locational limitations on work; Büscher 2014; Czarniawska 2014). In the following chapter, I discuss if and how these forms of flexible work support the satisfaction of (basic) psychological needs. In this context, I consider the need for autonomy, competence, and relatedness, which represent basic psychological needs according to self-determination theory (Ryan and Deci 2000). Furthermore, I consider the personal need for structure (Elovainio and Kivimäki 1998) as a need that is especially relevant in flexible working arrangements.

Previous research has found several advantages and disadvantages of flexible working arrangements, such as higher performance (Gajendran and Harrison 2007), work intensification (Kelliher and Anderson 2010), and work/non-work conflict (Allen et al. 2013). Integrating need satisfaction in this context may further deepen our understanding of the positive and negative effects of flexible working arrangements. In terms of practice, this discussion guides supervisors in designing

C. Gerdenitsch (✉)
Faculty of Psychology, Universitätsstraße 7, 1010 Vienna, Austria
e-mail: cornelia.gerdenitsch@univie.ac.at

C. Korunka and B. Kubicek (eds.), *Job Demands in a Changing World of Work*,
DOI 10.1007/978-3-319-54678-0_6

flexible working arrangements in a way that enables the satisfaction of subordinates' needs. In this chapter, I will particularly discuss research published over the last 25 years about need satisfaction in the context of flexible work. Google Scholar and Scopus were used as literature databases and searched with the following keywords: *need for autonomy, need for competence*, *need for relatedness*, and *need for structure* combined with each of the forms of flexible work considered here: *flex (i)time*, *flexplace/telecommuting/telework*, *nomadic work*. In addition, two general journals (*Journal of Applied Psychology* and *Journal of Organizational Behavior*) and two journals that focus on new forms of work (*New Technology Work and Employment* and *Computer Supported Cooperative Work: CSCW: An International Journal*) were searched for these keywords.

This chapter is structured as follows: First, to highlight the relevance of the issue of flexible work in Europe, recent data from the European Working Conditions Survey[1] about the prevalence of flexible work is presented and the three forms of flexible work are defined. In the second section, I describe the general nature of psychological needs with reference to need theories. I then discuss the satisfaction potential of the three forms of flexible work with regard to each of the four selected needs. In the last section, I present guidelines for designing flexible work arrangements with implications for leadership behavior. In this vein, the chapter seeks to assist practitioners in maximizing the benefits of flexible working arrangements and using them to their fullest potential.

6.2 Forms of Flexible Work in Europe

Working flexibly in terms of time and location is enabled by various developments in information and communication technologies, such as the enhanced storage capacity of batteries, the mobility of computers, wireless and broadband technologies, cloud-based services, and the pervasiveness of mobile devices. Recent data from the European Working Conditions Survey show that there are indeed a considerable number of flexible workers in Europe. The European Working Conditions Survey is conducted every five years to collect data about working conditions in Europe. In 2015, workers from the 28 European Union member states, Norway, Switzerland, Albania, the former Yugoslav Republic of Macedonia, Montenegro, Serbia, and Turkey were asked about their working conditions. On average, 39% of workers reported having the ability to select the times they start and finish working, with Finland, the Netherlands, and Austria having the highest prevalence (see Fig. 6.1). Furthermore, on average, 30% of workers reported that they worked from multiple locations.

Various forms arise from combining the temporal and locational components of flexible work. In this chapter, research findings and recommendations regarding

[1] http://www.eurofound.europa.eu/surveys.

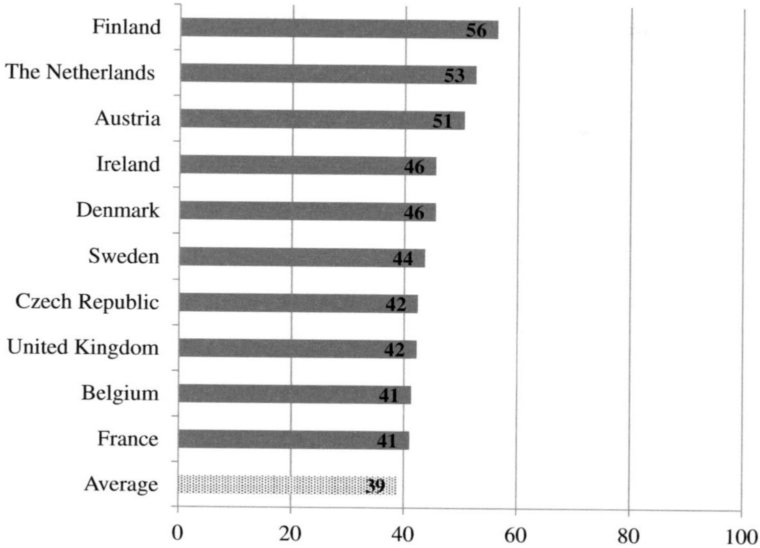

Fig. 6.1 Ten European countries with the highest temporal flexibility rates compared to the average in percentages (Eurofound 2015; http://www.eurofound.europa.eu/surveys)

three distinct forms of flexible work are presented, namely flextime (medium in temporal and low in locational flexibility), flexplace (medium in temporal and high in locational flexibility), and nomadic work (very high in temporal and locational flexibility). Flextime, or flexibility regarding time schedules, gives workers (some) choice over when and for how long work is performed. For instance, workers can decide when to begin and end work, but can also interrupt their work to do private chores (Hill et al. 2008). In flextime arrangements, core hours (e.g., from 10 a.m. to 3 p.m.) are usually established during which all workers have to be present.

Flexplace describes flexibility regarding work locations. Workers with locational flexibility may work at home, coffee shops, satellite offices, branch offices, remote working centers, coworking spaces, or while travelling or commuting in trains or airplanes, airports, and hotels (Bailey and Kurland 2002). Flexplace arrangements include teleworking or telecommuting. Teleworking means that workers work from somewhere outside the office on a regular basis (Sullivan 2003). Teleworking is often used interchangeably with telecommuting, which is also known as remote working (working from home), and represents a form of teleworking. Examples of flexplace workers are ones who work from home one day of the week (e.g., at Fridays), consultants who work at a client's office from Monday to Thursday, or a journalist who partially works on the road away from his or her agency.

Nomadic work includes both components of flexible work and thus lacks temporal and locational limitations (Kleinrock 1996). In contrast to workers in flextime and flexplace arrangements, nomadic workers do usually not have a regular office. Nomadic workers are anytime, anywhere workers who rely considerably on technologies to do their work (Büscher 2014; Czarniawska 2014). Because of this extensive use of technologies, these workers are also labelled digital nomads. Typical examples are freelancers working in the area of web design, programming, or language editing. However, non-digital nomadic jobs also exist, such as tour guides or hairstylists who visit clients at home, events, or in hotels.

There is no strict distinction between these three forms of flexible work. Workers can use all three forms of work from time to time, with one being the most dominant. Their common difference from traditional work arrangements regards the increasing degree of freedom to autonomously decide when and where to work (Hill et al. 2008). Thus, using forms of flexible work should satisfy workers' need for autonomy. But is this really the case? Do these forms of flexible work foster a (social) environment that can satisfy psychological needs for relatedness, competence, and structure as well? The next subchapter provides answers to these questions.

6.3 Need Satisfaction in the Context of Flextime, Flexplace, and Nomadic Work

Psychological needs received much attention in academic research during the 20th century (Baumeister and Leary 1995; Freud 1920; Maslow 1954; McDougall 1908; Murray 1938; Reis et al. 2000). Self-determination theory (SDT; Ryan and Deci 2000) identifies three of them as basic ones (see also Sheldon et al. 2001): autonomy, relatedness, and competence. The need for autonomy means that individuals want to have a sense of choice regarding their activities. The need for relatedness implies that individuals need to experience interpersonal connections with others. Finally, the need for competence relates to whether one's own activities are perceived as effective. In addition to these three needs, the need for structure describes a personal preference for structure and predictability in life (Neuberg and Newsom 1993). In flexible working arrangements, predefined temporal and locational structures (as they exist in traditional working arrangements) are broken up at least to some extent. Therefore, the need for structure is especially relevant in the context of flexible work (e.g., Van Yperen et al. 2014).

SDT states that psychological needs – like physiological needs (Hull 1943) – are innate organismic necessities (in contrast to, e.g., Murray 1938, who proposed that needs are acquired). In particular, SDT defines needs as "innate psychological nutriments that are essential for on-going psychological growth, integrity, and well-being" (Deci and Ryan 2000: 229). In other words, needs exist within each individual. SDT further takes into account differences regarding the degree to which need satisfaction is experienced. Thereby, SDT does not focus on the variety in need strengths, but on the degree to which individuals experience need satisfaction in different contexts (Deci and Ryan 2000). In contrast, the personal need for structure explicitly differs among individuals. In particular, it is assumed that

Forms of Flexible Work	Psychological Needs: Autonomy	Competence	Relatedness	Structure
Flextime	• Tension between flexibility for the organization and autonomy for workers (Gerdenitsch et al. 2015, Kattenbach et al. 2010; Knuth 1998)	• Possibility to adapt the work schedule to requirements of a work activity • Beneficial effects of flexible working on productivity, performance (Baltes et al. 1999), and a sense of mastery (Thomas and Gangster 1995)	• Regular physical encounters are possible, routinely ones are reduced	• Creating a structure (routines) requires special skills such as self-leadership, self-management, and self-control
Flexplace	• Beneficial effects are partially mediated by perceived autonomy (Gajendran and Harrison, 2007) • Critical role of managerial control mechanisms (e.g., electronic performance monitoring, Bhave 2013)	• Possibility to adapt the work location to requirements of a work activity • Depends on work activity and task complexity (see e.g., social facilitation theory, Zajonc 1965)	• Professional and social isolation with frequency of telecommuting as moderator (Bailey and Kurland, 2002) • Communication-enhancing technologies buffer negative effects of isolation (Golden et al. 2008), as does organizational social support (Bentley et al. 2016)	• Cues and rites facilitate the balance between one's desire for flexibility and the need for structure (Fonner and Stache 2012)
Nomadic Work	• Time and location independent work is positively related to a satisfied need for autonomy (van Yperen et al. 2014) • Unlimited autonomy within the context of the actual working task (Harmer and Pauleen 2012) • Autonomy paradox (Mazmanian et al. 2013)	• Possibility to to choose their work environment based on the requirements of their work and thus can master tasks more efficiently • Master 'setting up' their work environments	• Sense of isolation (Chen and Nath 2005; Harmer and Pauleen 2012) • Problems with team cohesion (Chen and Nath 2005) • Relevance of social networking tools or instant messengers (Harmer and Pauleen 2012)	• Defining structures self-determinated • Temporal and locational flexibility provokes ambiguity for workers with a high need for structure (Slijkhuis et al. 2013; van Yperen et al. 2014)

Fig. 6.2 Summary of findings about flextime, flexplace, and nomadic work in the context of psychological need satisfaction

individuals differ in their desire or personal preference for a simple structure (Neuberg and Newsom 1993).

Characteristics of the working environment can result in the frustration or fulfilment of need satisfaction (Deci and Ryan 2000). Flexible work greatly affects characteristics of the work environment and thus has the potential to influence need satisfaction. In the following four sections, I present scientific studies and arguments about the potential of flextime, flexplace, and nomadic work arrangements to promote need satisfaction. Within this, I focus on each of the selected needs (autonomy, competence, relatedness, structure) separately. A summary of the argumentation is provided in Fig. 6.2.

6.3.1 *Need for Autonomy*

The need for autonomy or independence describes a preference for "feeling like you are the cause of your own actions rather than feeling that external forces or pressures are the cause of your actions" (Sheldon et al. 2001; p. 339). In the working context, different dimensions of autonomy have been proposed, including autonomy regarding work methods, the work schedule, work time, and work place (De Spiegelaere et al. 2016). Several studies have already shown beneficial effects of perceiving autonomy in the working context. For instance, having the autonomy to control work time has a positive effect on productivity, performance, and job satisfaction (Baltes et al. 1999). Autonomy in the choice of work methods has been found to be beneficial for performance (Langfred and Moye 2004). The question remains whether work designs that provide some form of autonomy – as the types of flexible work do – enable satisfaction of the need for autonomy. Next, I will discuss how and when flextime, flexplace, and nomadic work arrangements can facilitate need satisfaction regarding autonomy.

Flextime. In flextime work arrangements, workers are in control of their work schedules (deciding when to start and quit working, as well as whether to interrupt work), and thus the perception of autonomy in terms of work-time control is higher than in non-flexible working arrangements. Knauth (1998) describes that traditional work arrangements provide low flexibility for both the organization and the worker. Instead, flextime arrangements provide a high degree of flexibility for workers (Knauth 1998). However, the need for autonomy can only be satisfied when the decision to work flexibly is indeed self-determined and not required by the organization or the work itself. Despite the expected increase in autonomy, choices about when and where to work can simultaneously be externally restricted by members of the organization or others with whom workers collaborate. Studies have found that flexible workers experience expectations of permanent availability from supervisors, clients, or colleagues (e.g., Gerdenitsch et al. 2015; Kattenbach et al. 2010). Furthermore, Kelliher and Anderson (2010) describe various types of work intensification that occur in flexible working arrangements. Thus, on the one hand, flextime can facilitate satisfaction of the need for autonomy, but, on the other hand, can also lead to external restrictions that prevent the perception of increased autonomy.

Flexplace. Previous studies on telecommuting have found beneficial effects on performance and absenteeism (Stavrou 2005). In a meta-analysis, Gajendran and Harrison (2007) were able to show that beneficial effects are at least partially mediated by perceived autonomy. In line with this, a study examining autonomy among teleworkers and those who stayed in the office (non-teleworkers) found higher job autonomy among teleworkers (O'Neill et al. 2009). The authors concluded, due to the cross-sectional nature of the study, that it is not clear whether more autonomous jobs are better suited for telework or whether teleworkers feel more autonomous because supervisors do not directly monitor them.

Managing teleworkers represents a challenge for supervisors (Kurland and Cooper 2002). Because behavior control, i.e. control by physical presence, is no longer possible, alternative forms of managerial control, such as electronic performance monitoring (Bhave 2013), output control, or normative control are increasingly applied. Electronic performance monitoring includes controlling employees by electronically collecting data about their work behaviors (e.g., GPS data, log files). Output control is comprised of control mechanisms like results-only work environments or management by objectives/results. Normative control represents controlling workers through a set of informal rules about working behavior. These norms represent scripts about how to act in social and work situations [for a detailed description of managerial control types, see Cooper and Kurland (2002) or

Ouchi and Johnson (1978)]. Depending on the control type, the potential to satisfy the need for autonomy can be strengthened (in the cases of output and clan control) or hindered (in the case of electronic performance monitoring). Thus, on the one hand, flexplace represents a work environment that can support satisfaction of the need for autonomy. On the other hand, a feeling of surveillance can arise because of the use of certain managerial control types, which in turn prevents satisfaction of the need for autonomy.

Nomadic work. Naturally, nomadic work is the form of flexible work that provides the highest levels of autonomy. Nomadic workers can fully decide when and where to do their work without any temporal and spatial limitations. In line with this, time and location independent work is found to be positively related to the satisfaction of people's need for autonomy (Van Yperen et al. 2014). Harmer and Pauleen (2012) point out that the apparently unlimited autonomy of nomadic work exists only within the context of the working task, which itself involves temporal and locational restrictions. For example, a freelance journalist would have to adapt his/her work schedule to the temporal and locational requirements of the persons he/she interviews. The same applies to a freelance hairstylist when working for a fashion show, for example. When it comes to collaboration, nomadic workers also have to adapt to co-workers' preferences. Especially in a globalized working world, co-workers are increasingly spread over different time zones, which necessitates a joint agreement about when to work. In such cases, the personal preferences of one nomadic worker may conflict with the requirements and preferences of his/her co-workers.

Another relevant aspect involves external expectations of the nomadic worker's availability. For example, each member of an international team working in the field of online payment solutions expects a certain degree of availability from their supervisor. This results in a norm of constant availability and, as a consequence, reduces the supervisor's ability to disconnect from work. In their recent article, Mazmanian et al. (2013) term this phenomenon the 'autonomy paradox'. More concretely, the autonomy paradox describes the fact that knowledge workers using mobile technologies to manage communication experience tension between personal autonomy and commitment to colleagues or clients.

In conclusion, forms of flexible work have a significant potential to satisfy the need for autonomy. This work design gives workers the autonomy to decide when and where to work. However, there is evidence that all three mentioned forms of flexible work also represent the proverbial double-edged sword, as workers' experienced autonomy is diminished by the flexibility demands of the organization, managerial control types, task limitations, and co-workers'/subordinates' expectations. Therefore, the potential to satisfy the need for autonomy depends on the combination of perceived autonomy and external limitations.

6.3.2 Need for Competence

Need for competence is a preference for "feeling that you are very capable and effective in your actions rather than feeling incompetent or ineffective" (Sheldon et al. 2001; p. 339). When the need for competence is satisfied in the context of work, workers feel that they have mastered the tasks required by their jobs, are good at these tasks, or can accomplish very difficult tasks (see the work-related need satisfaction scale by van den Broeck et al. 2013).

Flextime. In flextime arrangements, workers are able to adapt their working schedule to the requirements of their current task. In doing so, they can achieve results more efficiently or with higher quality than in non-flexible arrangements. In line with this, previous studies have identified beneficial effects of flexible work on productivity, performance (Baltes et al. 1999), and a sense of mastery (Thomas and Ganster 1995). Especially when working in geographically disparate teams (e.g., in different time zones), adaptation of the work schedule is necessary. Fixed schedules would hamper synchronic (virtual) communication, since videoconferencing across time zones requires careful organization and high flexibility for those involved. For example, a team with members in three different time zones might be having a complex problem with the product they are selling. In this case, a videoconference is needed to work on a solution to the problem. For everyone to join in on the conference call, each worker has to adapt their work schedule. A strict work schedule would hinder communication and therefore complicate and delay the solution to the problem.

Another scenario that demonstrates the limitations of fixed work schedules concerns workflow. Imagine a worker whose workday ends at 5 p.m. every day. If this worker finishes one task at 4:30 p.m., they either cannot start working on a new complex task or will have to interrupt his/her workflow at 5 p.m. This will negatively affect the efficiency and quality of the working task's output, and the workers' need for competence will only partially be satisfied.

Flexplace. Previous studies have found that telecommuting has positive effects on performance (Gajendran and Harrison 2007). Telecommuting allows workers to concentrate on their work in an undisturbed working environment (as long as they are not distracted by family issues; for an overview of telecommuting and work-family conflict, see Golden et al. 2006). Flexplace workers can more easily adapt the location of work to the requirements of their work activity. When they have to do work requiring strong concentration, they can seek an undisturbed private working environment, for example at home. When they have to do creative work, they can go to a pleasant, inspiring work environment. For collaborative activities, such as brainstorming, discussing problems or creating business strategies, work environments where co-workers are physically present will be more suitable. Furthermore, task complexity is relevant for flexplace workers. Easy and familiar tasks are performed better in the presence of others, while performance on difficult or unknown tasks is reduced (social facilitation theory; Zajonc 1965). Thus, doing familiar tasks in a coffee shop in the presence of other people could increase performance, while complex tasks may be harder to accomplish.

In general, when the work environment fits the needs of the specific work activity, workers may accomplish better results and thus feel more competent. However, despite these beneficial effects, there are also some problems with flexplace arrangements. In flexplace arrangements, direct verbal feedback is expected to be limited, which can be a relevant factor in experiencing a sense of competence. For example, consider a consultant communicating with a client via mobile phone while a colleague is sitting next to him/her. This colleague can listen and provide direct feedback about communication techniques. This is especially important for new employees who are still unsure about how to communicate with clients correctly. In such a case, the physical presence of a colleague who can listen and provide direct feedback can contribute to an increased feeling of competence.

Nomadic work. Nomadic workers can experience a feeling of competence not only in the context of performing their work tasks but also in managing their workspace. First, nomadic workers are free to choose their work environment based on the requirements of their work and thus can master tasks more efficiently. In other words, they are expected to experience the highest congruence between the needs of the current work activity and the amenities of the work environment, which they create themselves. However, empirical support showing that temporal and locational flexibility increase satisfaction of the need for competence is still lacking (Van Yperen et al. 2014). Second, nomadic workers can feel competent when they are able to organize their workflow efficiently and successfully. In particular, they have to master 'setting up' their work environments. This includes organizing their physical (e.g., paper-based materials) and digital (e.g., a suitable cloud system and software equipment) working environment.

To sum up, flexible forms of work can help workers feel competent in mastering their working tasks as they can adapt their working schedule and work location to the requirements of their current work activities. Due to the congruence that can be created between work activities and the work environment, high satisfaction of the basic psychological need for competence is expected. Furthermore, successfully setting up the work environment should contribute to need satisfaction.

6.3.3 Need for Relatedness

The need for relatedness is a preference for "feeling that you have regular intimate contact with people who care about you rather than feeling lonely and uncared for" (Sheldon et al. 2001, p. 339). It refers to the desire to feel connected to others (Deci and Ryan 2000). According to van den Broeck et al. (2010), the need to integrate oneself into a social context and benefit from it is consistent with the concept of social support. Social support includes the exchange of information, direct support, and emotional support such as admiration and liking or confirmation (House et al. 1988; Kahn and Antonucci 1980). Social support in the working context is mainly provided by colleagues and supervisors in the form of feedback, listening to problems, or helping with work tasks (Baruch-Feldman et al. 2002).

Flextime. Because of core hours in flextime arrangements, people still meet each other in the office. Thus, a environment with the potential to satisfy the need for relatedness exists. However, routine physical encounters such as leaving the office with the same co-workers every day or meeting one another at the coffee machine regularly may be more common when starting and finishing times are predefined and the same for all. Nevertheless, in flextime arrangements, regular physical encounters can be more common with workers with similar daily routines.

Flexplace. Studies on telework point out difficulties in terms of social and professional isolation, especially when workers do not have access to informal developmental opportunities such as mentoring, interpersonal networking, or informal learning (Cooper and Kurland 2002) have little impact on how often people communicate (Duxbury and Neufeld 1999). Also, the non-teleworker to teleworker ratio is relevant here (Golden 2007). For example, in a team in which all members use teleworking from time to time, a single member will feel less isolated than if he/she is the only one who teleworks. Negative effects of isolation can be reduced via face-to-face interactions, access to communication-enhancing technologies (Golden et al. 2008), and organizational support for low-intensity teleworkers (Bentley et al. 2016).

Conversely, instead of feeling isolated, workers may experience an increased relatedness with colleagues. Teleworkers have to use various computer-mediated communication technologies to collaborate and communicate with co-workers. These technologies provide broad possibilities to get in contact with each other. Introverts, for instance, may get in contact with others more easily compared to face-to-face contact, since they may feel more comfortable using computer-mediated communication technologies.

Nomadic work. Studies on anytime anywhere workers show that they perceive a sense of isolation from their co-workers (Chen and Nath 2005; Harmer and Pauleen 2012), leading to lower team cohesiveness as a consequence (Chen and Nath 2005). Both studies conclude that communication technologies such as social networking tools or instant messengers should help workers avoid feelings of isolation. Thus, organizations should provide technologies and create a norm that their workers integrate them into their daily communication habits and provide cultural support for the use of ubiquitous computing technologies (Chen and Nath 2005). Freelancers and self-employed will benefit from visiting Coworking spaces. Coworking spaces are shared office environments providing a basic business infrastructure (Spinuzzi 2012) as well as possibilities for social interaction (Gerdenitsch et al. 2016).

To summarize, for the feeling of relatedness, regular physical or virtual encounters play a significant role. Regular encounters at work may be easier when workers are present at the same time and in the same place. However, computer-mediated communication technologies enable a vast number of communication methods, including synchronous (videoconferencing) and asynchronous communication (email, messenger). Using these technologies can buffer the negative effects of isolation and even increase a sense of relatedness.

6.3.4 Need for Structure

In addition to the three basic psychological needs for autonomy, competence, and relatedness (Deci and Ryan 2000), the need for structure is especially relevant in the context of flexible work. The need for structure describes a personal preference for structure and predictability in life (Neuberg and Newsom 1993). Individuals with a high need for structure experience discomfort in situations that lack clarity and structure. They then start establishing routines or rely on formalized social scripts to increase structure, which at the same time reduces cognitive load (Neuberg and Newsom 1993; Thompson et al. 2001). Research on the personal need for structure identifies two dimensions: first, the individual's desire or affinity for a structured environment, and second, their response to a lack of structure in unstructured and unpredictable environments (Elovainio and Kivimäki 1998). The desire for structure can function as a job resource, whereas a lack of structure can lead to higher risks of psychological strain in highly complex jobs (Elovainio and Kivimäki 1998).

Flextime. In flextime arrangements, temporal work structures are broken up to some extent, which decreases opportunities to satisfy the need for structure. However, because flextime arrangements provide the opportunity for workers to decide when to start and finish working, workers can choose to maintain a fixed working schedule. Nevertheless, those workers who want structure in flextime arrangements have to build it on their own. Sticking to these structures may cost energy and require specific skills such as self-leadership (for an overview see Neck and Houghton 2006), self-management, or self-control (Baumeister et al. 1994). Thus, flextime arrangements can fulfill the desire for a structured environment if flexible workers are able to build these structures themselves.

Flexplace. In the context of flexplace, workers can also structure their work in such a way that the need for structure is satisfied. Research shows that teleworkers use cues and rites to facilitate the balance between their need for flexibility and their need for structure (Fonner and Stache 2012). The authors found that locationally flexible workers set physical barriers when shifting from a work to a personal role. For instance, they report assigning personal and work roles to different spaces and actively leaving or entering these spaces. Closing the door of the room one is working in can also function to separate work time from personal time.

Nomadic work. For nomadic workers, there is no predefined temporal or locational structure. They have to structure their work activities, organize their settings, define boundaries between work and non-work, and organize the technologies they need for work on their own. Such self-defined work environments can be created in such a way that the need for structure is fulfilled. However, defining structures and sticking to them without support from others is expected to cost energy (see Prem, Chap. 7). Using coworking spaces as flexible office spaces can be helpful in giving nomadic work more structure.

Individuals will appraise nomadic work differently depending on the strength of their need for structure. Previous studies indicate that workers with a high need for structure have problems with temporal and locational flexibility because it provokes ambiguity (Slijkhuis et al. 2013; Van Yperen et al. 2014). Ambiguity causes workers to be unsure about expectations at work and makes them feel that they cannot anticipate future events (Elovaino and Kivimäki 2001). Those who experience ambiguity and have a high need for structure will behave in a way to reduce ambiguity. For example, they will structure their day or week in such a way that they have enough structure.

To sum up, forms of flexible work reduce or remove a predefined structure of when and where to work. Workers who desire a structured work environment may experience ambiguity because of unstructured and unpredictable situations. They will then create desired structures for and by themselves to make themselves feel comfortable. In doing so, and when they manage to stick to these structures, the need for structure can be satisfied.

6.3.5 Need Interdependencies

In contrast to the three basic psychological needs (autonomy, competence, relatedness), which according to Deci and Ryan (2000) exist within each individual with no regard to variety in need strengths, the strength of the need for structure explicitly differs among individuals (Elovainio and Kivimäki 2001). Thus, the personal need for structure may vary across different members of a work group. Let us assume a supervisor leads a work group with seven subordinates, where one has a very strong and another a very weak personal need for structure. Let us further assume three scenarios: One in which the supervisors provides the possibility for *flextime* with core hours from 11 a.m. to 2 p.m., one with the possibility of working at maximum three days from home or from a third place (*flexplace*), and one where there is no fixed time and place of work (*nomadic work arrangement*).

I now assume that workers will differ in the way they use the provided work arrangement. In particular, the team member with a strong preference for structure will behave differently from the one with a weak preference for structure. For example, the first worker will use the provided flexibility less intensively than the worker with a low need for structure. The former will also apply routines and/or establish rules to structure the day. I argue that these expected different ways of behaving will influence the degree to which the remaining psychological needs under consideration (autonomy, competence, relatedness) are satisfied. The following figure paints a broad picture about the extent to which Worker A with a high need for structure (solid line) is expected to experience need satisfaction in contrast to Worker B with a weak need for structure (dotted line) (Fig. 6.3).

I have argued that the potential to satisfy the need for autonomy in terms of choosing when and where to work is highest in nomadic work followed by flexplace and flextime arrangements. Worker A may not use the provided autonomy as intensively as Worker B in any of the three arrangements compared. In particular, Worker A will define routine starting and ending times for working, for example. Other team members will then also start to expect these recurring work patterns, which may lead to the emergence of external expectations about presence. Thus, the

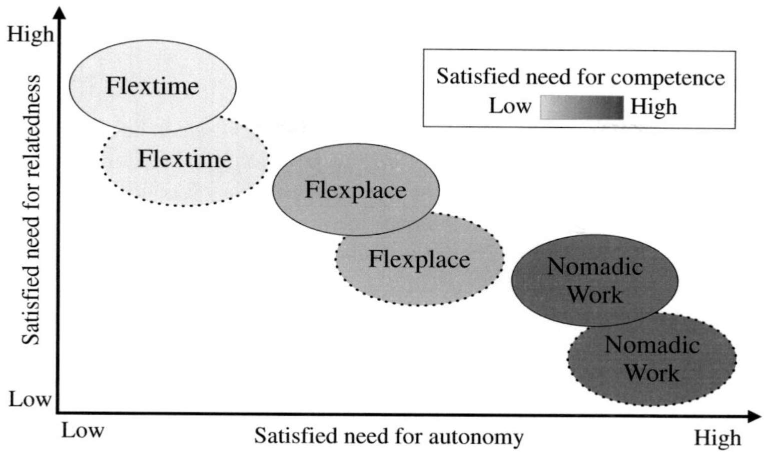

Fig. 6.3 Need satisfaction provided by the three forms of flexible work separated for people with strong (*solid line*, Worker A) and weak (*dotted line*, Worker B) need for structure

need for autonomy is expected to be less satisfied for Worker A compared to Worker B.

In terms of relatedness, I have argued that flextime has the highest potential for need satisfaction, assuming that communication technologies are not used extensively. Compared to flexplace and nomadic work, flextime arrangements provide better conditions for high physical proximity between co-workers. I again assume differences between Worker A and Worker B. Worker A may establish various daily routines like planned breaks and could thus for example, be met at specific places (e.g., office coffee machine, coworking space, café) at the same time each day. The resulting high predictability of Worker A's behavior will facilitate regular contacts.

Finally, in terms of competence, nomadic workers have the greatest freedom in creating or setting up a work environment suitable to successfully performing work activities, under the assumption that workers have good skills in managing their work environment. Thus, the need for competence may be most likely to be fulfilled in this form of flexible work. Worker A can be expected to create an environment with fixed boundaries, while Worker B will create different environments across days and weeks. Whether this behavior supports satisfaction of the need for competence depends on many factors such as task complexity, personal preferences, and type of work activity. I thus do not expect main differences between Worker A and Worker B in need satisfaction of competence here.

6.4 Designing Flexible Work to Support Need Satisfaction – Leadership Implications

To maintain psychological health, all basic psychological needs have to be satisfied (Deci and Ryan 2000). However, not only is psychological health influenced by need satisfaction, but fulfilment of basic needs is also a precondition for intrinsic motivation (Gagné and Deci 2005), which itself is a key predictor of performance, innovation, and productivity. The next section will provide some guidelines for supervisors about what to consider when designing flexible working arrangements for their team(s) or subordinate(s).

Need for autonomy in the context of flexible work can be fulfilled when workers have the ability to control their work schedule and location autonomously. Three important aspects should be considered. First, giving employees autonomy about their work must also include autonomy about when to stop working. Kelliher and Anderson (2010) have found work intensification to be among the consequences of flexible work, and it is crucial to avoid unnoticed exhaustion.

Second, managerial control types have the potential to both strengthen and mitigate satisfaction of the need for autonomy. For instance, using electronic performance monitoring can lead to a perception of surveillance and consequently inhibit satisfaction of the need for autonomy. Supervisors are challenged to find a balance between providing autonomy and monitoring staff. For some working tasks (e.g., monotonous repetitive tasks), electronic performance monitoring may be suitable and can even motivate workers to increase their performance. On the other hand, electronic performance monitoring is ill-suited to solving complex problems. A combination of monitoring systems that is tailored to the task will be most beneficial here. Leaders can also let their employees participate in the design and implementation of personalized control mechanisms. These control mechanisms can then be adapted over time based on evaluations from both the perspective of both leaders and subordinates.

Finally, when some team members are provided with autonomy while others are not, conflicts or fairness issues may arise. Golden (2007) found decreased job satisfaction rates for non-teleworkers in organizations in which teleworking was salient. Those with no autonomy may be even less satisfied when others are provided with autonomy. Therefore, it is important to be aware that providing autonomy for some workers affects those who are not provided with it. To avoid future conflicts, supervisors should either offer flexible work arrangements to all workers or comprehensibly communicate decisions about why certain positions have flexibility and others do not.

Need for competence can be satisfied when workers are able to master their working tasks successfully. This is facilitated when workers have the ability to identify a suitable and supportive working environment for their current tasks, an ability that should be encouraged and supported by leaders. At the same time, supervisors must accept that workers are not accessible while they are working in a private work environment. If workers have problems completing their tasks on their own, supervisors should provide (virtual) channels to their employees to collect feedback or input from others.

Need for relatedness can also be very actively satisfied by flexible forms of work, as long as forms of flexibility are designed in such a way that they address potential feelings of isolation or being out of sight. Employers should establish computer-mediated communication channels among team members so that those who are in the office and those who are not have a continued and regular exchange. In addition, these channels should be used to confirm that all important information is distributed to all workers. Furthermore, regular physical meetings between team members as well as informal work activities can increase team cohesion and identification with the team. Finally, it is recommended to establish activities that create a norm of social support among all team members no matter when and where they are working.

Need for structure. Two things are important to support satisfaction of the personal need for structure: providing a structure and supporting workers in developing skills to manage flexible work. A work structure can include predefined meeting times (e.g., weekly team meeting), a definition of core hours when workers should be at least virtually available, and rules about the possibility of using coworking spaces and other shared spaces. In addition, supervisors should support their workers in developing skills and strategies to deal with the tension between flexibility and structure. This may include training to develop self-leadership strategies and support in setting and establishing routines. Leaders can also motivate their subordinates to test different techniques or foster exchange among team members about how they structure their work.

In conclusion, besides the need to investigate how workers themselves can cope with work designs resulting from socio-economic developments (see Flecker et al., Chap. 2), there is at the same time the need to further investigate how to co-create these work designs in order to satisfy psychological needs. Indeed, there are various methods *leaders* can use to design flexible forms of work in a way that promotes the satisfaction of subordinates' needs for autonomy, competence, relatedness, and structure. Flexible work is a dominant trend in the world of work, and leaders should know how to unlock its full potential and maximize its benefits.

References

Allen, T. D., Johnson, R. C., Kiburz, K. M., & Shockley, K. M. (2013). Work-family conflict and flexible work arrangements: Deconstructing Flexibility. *Personnel Psychology, 66,* 345–376. doi:10.1111/peps.12012

Bailey, D. E., & Kurland, N. B. (2002). A review of telework research: Findings, new directions, and lessons for the study of modern work. *Journal of Organizational Behavior, 23,* 383–400. doi:10.1002/job.144

Baltes, B. B., Briggs, T. E., Huff, J. W., Wright, J. A., & Neuman, G. A. (1999). Flexible and compressed workweek schedules: A meta-analysis of their effects on work-related criteria. *Journal of Applied Psychology, 84,* 496–513. doi:10.1037/0021-9010.84.4.496

Baruch-Feldman, C., Brondolo, E., Ben-Dayan, D., & Schwartz, J. (2002). Sources of social support and burnout, job satisfaction, and productivity. *Journal of Occupational Health Psychology, 7,* 84–93. doi:10.1037//1076-8998.7.1.84

Baumeister, R. F., Heatherton, T. F., & Tice, (1994). *Losing control: Ho wand why people fail at self-regulation.* San Diego, CA: Academic Press.

Baumeister, R. F., & Leary, M. R. (1995). The need to belong: Desire for interpersonal attachments as a fundamental human motivation. *Psychological Bulletin, 117,* 497–529. doi:10.1037/0033-2909.117.3.497

Bentley, T. A., Teo, S. T. T., McLeod, L., Tan, F., Bosua, R., & Gloet, M. (2016). The role of organisational support in teleworker wellbeing: A socio-technical systems approach. *Applied Ergonomics, 52,* 207–215. doi:10.1016/j.apergo.2015.07.019

Bhave, D. P. (2013). The invisible eye? Electronic performance monitoring and employee job performance. *Personnel Psychology, 00,* 1–31. doi:10.1111/peps.12046

Büscher, M. (2014). Nomadic work: Romance and reality. A response to Barbara Czarniawska's 'nomadic work as life-story plot'. *Computer Supported Cooperative Work (CSCW), 23,* 223–238. doi:10.1007/s10606-013-9194-6

Chen, L., & Nath, R. (2005). Nomadic culture: Cultural support for working anytime, anywhere. *Information Systems Management, 22,* 56–64. doi:10.1201/1078.10580530/45520.22.4.20050901/90030.6

Cooper, C. D., & Kurland, N. B. (2002). Telecommuting, professional isolation, and employee development in public and private organizations. *Journal of Organizational Behavior, 23,* 511–532. doi:10.1002/job.145

Czarniawska, B. (2014). Nomadic work as life-story plot. *Computer Supported Cooperative Work (CSCW), 23,* 205–221. doi:10.1007/s10606-013-9189-3

De Spiegelaere, S., Van Gyes, G., & Van Hootegem, G. (2016). Not all autonomy is the same. Different dimensions of job autonomy and their relation to work engagement and innovative work behavior. *Human Factors and Ergonomics in Manufacturing & Service Industries, 0,* 1–13. doi:10.1002/hfm.20666

Deci, E. L., & Ryan, R. M. (2000). The "what" and " why" of goal pursuits: Human needs and the self-determination of behavior. *Psychological Inquiry, 11,* 227–268. doi:10.1207/S15327965PLI1104_01

Demerouti, E., Derks, D., Ten Brummelhuis, L. L., & Bakker, A. B. (2014). New ways of working: Impact on working conditions, work-family balance, and well-being. In C. Korunka & P. Hoonakker (Eds.), *The impact of ICT on quality of working life* (pp. 123–142). New York, Amsterdam: Springer.

Duxbury, L., & Neufeld, D. (1999). An empirical evaluation of the impacts of telecommuting on intra-organizational communication. *Journal of Engineering and Technology Management, 16,* 1–28. doi:10.1016/S0923-4748(98)00026-5

Elovainio, M., & Kivimäki, M. (1998). Personal need for structure and occupational strain: An investigation of structural models and interaction with job complexity. *Personality and Individual Differences, 26,* 209–222. doi:10.1016/S0191-8869(98)00038-5

Elovainio, M., & Kivimäki, M. (2001). The effects of personal need for structure and occupational identity in the role stress process. *The Journal of Social Psychology, 141,* 365–378. doi:10.1080/00224540109600558

Eurofound. (2015). First findings: Sixth European Working Conditions Survey. doi:10.2806/59106

Fonner, K. L., & Stache, L. C. (2012). All in a day's work, at home: Teleworkers' management of micro role transitions and the work–home boundary. *New Technology, Work and Employment, 27,* 242–257. doi:10.1111/j.1468-005X.2012.00290.x

Freud, S. (1920). *Beyond the pleasure principle.* London: Hogarth.

Gagné, M., & Deci, E. L. (2005). Self-determination theory and work motivation. *Journal of Organizational Behavior, 26,* 331–362. doi:10.1002/job.322

Gajendran, R. S., & Harrison, D. A. (2007). The good, the bad, and the unknown about telecommuting: Meta-analysis of psychological mediators and individual consequences. *Journal of Applied Psychology, 92,* 1524–1541. doi:10.1037/0021-9010.92.6.1524

Gerdenitsch, C., Kubicek, B., & Korunka, C. (2015). Control in flexible working arrangements: When freedom becomes duty. *Journal of Personnel Psychology, 14,* 61–69. doi:10.1027/1866-5888/a000121

Gerdenitsch, C., Scheel, T. E., Andorfer, J., & Korunka, C. (2016). Coworking spaces: A source of social support for independent professionals. *Frontiers in Psychology, 7.* doi:10.3389/fpsyg.2016.00581

Golden, T. (2007). Co-workers who telework and the impact on those in the office: Understanding the implications of virtual work for co-worker satisfaction and turnover intentions. *Human Relations, 60,* 1641–1667. doi:10.1177/0018726707084303

Golden, T. D., Veiga, J. F., & Dino, R. N. (2008). The impact of professional isolation on teleworker job performance and turnover intentions: Does time spent teleworking, interacting face-to-face, or having access to communication-enhancing technology matter? *Journal of Applied Psychology, 93,* 1412–1421. doi:10.1037/a0012722

Golden, T. D., Veiga, J. F., & Simsek, Z. (2006). Telecommuting's differential impact on work-family conflict: Is there no place like home? *Journal of Applied Psychology, 91,* 1340–1350. doi:10.1037/0021-9010.91.6.1340

Harmer, B. M., & Pauleen, D. J. (2012). Attitude, aptitude, ability and autonomy: The emergence of "offroaders", a special class of nomadic worker. *Behaviour & Information Technology, 31,* 439–451. doi:10.1080/0144929X.2010.489117

Hill, J. E., Grzywacz, J. G., Allen, S., Blanchard, V. L., Matz-Costa, C., Shulkin, S., et al. (2008). Defining and conceptualizing workplace flexibility. *Community, Work & Family, 11,* 149–163. doi:10.1080/13668800802024678

House, J. S., Umberson, D., & Landis, K. R. (1988). Structure and process of social support. *Annual Review Sociology, 14,* 293–318. doi:10.1146/annurev.so.14.080188.001453.

Hull, C. L. (1943). *Principles of behavior: An introduction to behavior theory.* New York: Appleton-Century-Crofts.

Kahn, R. L., & Antonucci, T. (1980). Convoys over the life course: attachment, roles and social support. In P. B. Baltes & O. Brim (Eds.), *Life span development and behavior* (pp. 254-283). Boston: Lextington Press.

Kattenbach, R., Demerouti, E., & Nachreiner, F. (2010). Flexible working times: Effects on employees' exhaustion, work-nonwork conflict and job performance. *Career Development International, 15,* 279–295. doi:10.1108/13620431011053749

Kelliher, C., & Anderson, D. (2010). Doing more with less? Flexible working practices and the intensification of work. *Human Relations, 63,* 83–106. doi:10.1177/0018726709349199

Kleinrock. (1996). Nomadicity anytime, anywhere in a disconnected world. *Mobile Networks and Applications, 1,* 351–357. http://dl.acm.org/citation.cfm?id=274618

Knauth, P. (1998). Innovative worktime arrangements. *Scandinavian Journal of Work, Environment & Health, 3,* 13–17. http://www.sjweh.fi/show_abstract.php?abstract_id=330

Kurland, N. B., & Cooper, C. D. (2002). Manager control and employee isolation in telecommuting environments. *The Journal of High Technology Management Research, 13,* 107–126. doi:10.1016/S1047-8310(01)00051-7

Langfred, C. W., & Moye, N. A. (2004). Effects of task autonomy on performance: An extended model considering motivational, informational, and structural mechanisms. *Journal of Applied Psychology, 89,* 934–945. doi:10.1037/0021-9010.89.6.934

Maslow, A. H. (1954). *Motivation and personality.* New York: Harpers.

Mazmanian, M., Orlikowski, W. J., & Yates, J. (2013). The autonomy paradox: The implications of mobile email devices for knowledge professionals. *Organization Science, 24,* 1337–1357. doi:10.1287/orsc.1120.0806

McDougall, W. (1908). *Introduction to social psychology.* London: Methuen.

Murray, H. (1938). *Explorations in personality.* New York: Oxford University Press.

Neck, C. P., & Houghton, J. D. (2006). Two decades of self-leadership theory and research: Past developments, present trends, and future possibilities. *Journal of Managerial Psychology, 21,* 270–295. doi:10.1108/02683940610663097

Neuberg, S. L., & Newsom, J. T. (1993). Personal need for structure: Individual differences in the desire for simpler structure. *Journal of Personality and Social Psychology, 65,* 113–131. doi:10.1037/0022-3514.65.1.113

O'Neill, T. A., Hambley, L. A., Greidanus, N. S., MacDonnell, R., & Kline, T. J. (2009). Predicting teleworker success: An exploration of personality, motivational, situational, and job characteristics. *New Technology, Work and Employment, 24,* 144–162. doi:10.1111/j.1468-005X.2009.00225.x

Ouchi, W. G., & Johnson, J. B. (1978). Types of organizational control and their relationship to emotional well being. *Administrative Science Quarterly, 23,* 293–317. doi:10.2307/2392566

Reis, H. T., Sheldon, K. M., Gable, S. L., Roscoe, J., & Ryan, R. M. (2000). Daily well-being: The role of autonomy, competence, and relatedness. *Personality and Social Psychology Bulletin, 26,* 419–435. doi:10.1177/0146167200266002

Ryan, R. M., & Deci, E. L. (2000). Self-determination theory and the facilitation of intrinsic motivation, social development, and well-being. *American Psychologist, 55,* 68–78. doi:10.1037/0003-066X.55.1.68

Sheldon, K. M., Elliot, A. J., Kim, Y., & Kasser, T. (2001). What is satisfying about satisfying events? Testing 10 candidate psychological needs. *Journal of Personality and Social Psychology, 80,* 325–339. doi:10.1037//O022-3514.80.2.325

Slijkhuis, J. M., Rietzschel, E. F., & Van Yperen, N. W. (2013). How evaluation and need for structure affect motivation and creativity. *European Journal of Work and Organizational Psychology, 22,* 15–25. doi:10.1080/1359432X.2011.626244

Spinuzzi, C. (2012). Working alone together: Coworking as emergent collaborative activity. *Journal of Business and Technical Communication, 26,* 399–441. doi:10.1177/1050651912444070

Stavrou, E. T. (2005). Flexible work bundles and organizational competitiveness: A cross-national study of the European work context. *Journal of Organizational Behavior, 26,* 923–947. doi:10.1002/job.356

Sullivan, C. (2003). What's in a name? Definitions and conceptualisations of teleworking and homeworking. *New Technology, Work and Employment, 18,* 158–165. doi:10.1111/1468-005X.00118

Thomas, L. T., & Ganster, D. C. (1995). Impact of family-supportive work variables on work-family conflict and strain: A control perspective. *Journal of Applied Psychology, 80,* 6–15. doi:10.1037//0021-9010.80.1.6

Thompson, M. M., Naccarato, M. E., Parker, K. C. H., & Moskowitz, G. (2001). The personal need for structure (PNS) and personal fear of invalidity (PFI) scales: Historical perspectives, present applications and future directions. In G. Moskowitz (Ed.), *Cognitive social psychology: The Princeton symposium on the legacy and future of social cognition* (pp. 19–39). Mahwah, NJ: Erlbaum.

van den Broeck, A., Lens, W., De Witte, H., & Van Coillie, H. (2013). Unraveling the importance of the quantity and the quality of workers' motivation for well-being: A person-centered perspective. *Journal of Vocational Behavior, 82,* 69–78. doi:10.1016/j.jvb.2012.11.005

van den Broeck, A., Vansteenkiste, M., Witte, H., Soenens, B., & Lens, W. (2010). Capturing autonomy, competence, and relatedness at work: Construction and initial validation of the work-related basic need satisfaction scale. *Journal of Occupational and Organizational Psychology, 83,* 981–1002. doi:10.1348/096317909X481382

Van Yperen, N. W., Rietzschel, E. F., & De Jonge, K. M. (2014). Blended working: For whom it may (not) work. *PLoS ONE, 9.* doi:10.1371/journal.pone.0102921

Zajonc, R. B. (1965). Social facilitation. *Science, 149,* 269–274. doi:10.1126/science.149.3681.269

Author Biography

Cornelia Gerdenitsch is a doctoral researcher at the Faculty of Psychology, University of Vienna. Her current research is embedded in the context of the digitalization of the workplace. In particular, she focuses on the effects of office environments (i.e., coworking spaces, activity-based flexible offices) on workers. She is further interested in the effects of flexible working arrangements on performance and well-being.

Chapter 7
The Effects of a Changing World of Work on Daily Working Life

Roman Prem

The world of work has changed considerably in recent decades as societal, economic, and technological changes driven by financialization, the network economy, and digitalization (see also Flecker et al., Chap. 2) how work is structured and carried out (Cascio 1995; Rosa 2003, 2013). In modern work settings, employees are faced with new job demands, which drain their resources and threaten their well-being (e.g., Kubicek et al. 2015). For example, work intensification puts additional demands on employees (see also Paškvan and Kubicek, Chap. 3) and in some cases job autonomy may even lose its beneficial effects and start to show its dark sides (see also Kubicek et al., Chap. 4). Further, a shift of responsibility for career management from organizations to employees themselves can also be considered a new demand. However, crafting one's career to be in line with one's goals (see also Nalis, Chap. 5) and using forms of flexible work to satisfy one's psychological needs (see also Gerdenitsch, Chap. 6) may also hold the potential for positive effects, as these aspects of modern work bring additional opportunities for growth, motivation, and learning. Overall, changing work environments appear to hold the potential for both adverse as well as beneficial effects on employees' well-being and performance in daily working life.

To explain how changes in modern work settings can affect daily working life, the present chapter will first give an overview of the state of research on the underlying mechanisms that link work characteristics to work outcomes. In particular, the present chapter will discuss the role of action regulation (e.g., Frese and Zapf 1994; Diestel and Schmidt 2012), cognitive appraisal (e.g., Lazarus and Folkman 1984; Searle and Auton 2015), and motivational processes (e.g., Ryan and Deci 2000; Gagné and Deci 2005) before integrating them into a framework model. Later on, I will discuss how changes in modern work settings might affect these

R. Prem (✉)
Faculty of Psychology, University of Vienna, Universitätsstraße 7,
1010 Vienna, Austria
e-mail: roman.prem@univie.ac.at

C. Korunka and B. Kubicek (eds.), *Job Demands in a Changing World of Work*,
DOI 10.1007/978-3-319-54678-0_7

within-person processes of action regulation, cognitive appraisal, and motivation. Finally, the focus will shift to the potentially ambivalent consequences of changes in modern work environments for daily work outcomes. Both adverse (e.g., job strain, burnout, ego depletion, procrastination) and beneficial (proactivity, job engagement, well-being, and thriving at work) consequences will be considered.

7.1 Linking Work Characteristics to Work Outcomes in Daily Working Life

To obtain more thorough knowledge about the short-term processes through which work characteristics affect work outcomes, researchers in work and organizational psychology have increasingly turned to diary studies over the past few decades (cf. Fisher and To 2012; Ohly et al. 2010). In contrast to cross-sectional studies that only allow researchers to analyze their data and interpret their results at the inter-individual or between-person level, diary studies enable researchers to investigate phenomena and processes at the intra-individual or within-person level.

Diary studies usually require participants to fill out short questionnaires once or even several times per workday over a period of one or more work weeks (e.g., Ohly et al. 2010). At each measurement occasion, participants may be asked to indicate their daily job demands and job resources and/or their momentary thoughts, feelings, and behaviors in naturalistic work settings. By "capturing life as it is lived" (Bolger et al. 2003: 579), diary methods reduce the probability of bias due to recall and retrospection that often arises in questionnaires. Even more importantly, repeating measures within comparably short periods of time allows researchers to investigate the within-person processes through which job demands and job resources impact work outcomes in daily working life.

Researchers in work and organizational psychology have investigated many kinds of within-person processes using diary study designs. Initially, research focused on the interplay between work and leisure time recovery (e.g., Sonnentag 2001; Sonnentag and Bayer 2005), which sparked streams of research focused on the effects of different recovery activities (Sonnentag and Fritz 2007) and the role of psychological detachment for leisure time recovery (Sonnentag and Fritz 2015). Despite the undisputed importance of research on psychological detachment and recovery during leisure time, the present chapter will focus on more immediate within-person processes that take place during the workday as work is carried out. Drawing on action regulation theory (e.g., Frese and Zapf 1994), the transactional theory of stress (e.g., Lazarus and Folkman 1984), and self-determination theory (Ryan and Deci 2000) as well as recent diary studies based on these theories, the following subsections of this chapter will give an overview of the within-person processes of action regulation, cognitive appraisal, and motivation.

7.1.1 Action Regulation Processes: Goal-Oriented Behavior and Self-Control Effort

Action regulation theory (e.g., Frese and Zapf 1994; Hacker 2003) describes within-person processes that guide work behavior. It defines *action* as goal-oriented behavior that can be regulated consciously or via routines and describes the action process across different levels of action regulation (Frese and Zapf 1994: 271).

At the beginning of the action process, work tasks that are assigned by the organization are usually translated into work goals. The translation of work tasks into work goals usually involves a process of redefinition (Hackman 1970), which is influenced by a number of factors, like the employee's understanding of the external task and its clarity. The resulting work goals can be described as anticipated future results that help guide the action process (Frese and Zapf 1994). In the next steps of the action process, employees generate action plans that take their evaluation of the environment into account and decide which action program to enact. Although the action programs generated may be more or less detailed, they are usually broken down into lower-level sub-programs that often relate to sub-goals of the more general work goal being pursued (Hacker 2003; Frese and Zapf 1994).

During the execution of the action programs and their sub-programs, employees use feedback to repeatedly compare their performance to the respective goals or sub-goals in order to assess their progress. The feedback used to monitor the execution of work programs can take many forms. For lower-level sub-programs, feedback often originates from the action itself or is available externally in the form of signals (e.g., error messages), whereas for higher-level programs feedback is also often provided by supervisors or co-workers. Once the feedback tells an employee that a certain sub-goal has been achieved, the employee will turn to the next sub-program and try to make progress on the next sub-goal (Hacker 2003; Frese and Zapf 1994).

Although goals are formed at a conscious level, they do not necessarily stay in consciousness over the entire action process. Instead, they are often only brought back into consciousness when (sub-)programs do not lead to the achievement of their associated (sub-)goals. Actions themselves can be regulated at three levels that differ in their requirements in terms of cognitive processing and effort (cf. Hacker 2003; Frese and Zapf 1994; for examples see also Zapf 2002). At the *intellectual level* of action regulation, action programs are designed for new work goals respectively adapted when they do not result in the achievement of the goals associated with them (e.g., because of changes in environmental conditions). At this level, processing consists of conscious problem solving and decisions to generate new strategies and plans. As a consequence, action regulation at the intellectual level is rather slow and resource-intensive compared to action regulation at other levels. At the *knowledge-based level*, employees rely on previously generated action programs, which are triggered by well-trained situational cues and signals. As a consequence, action regulation at this level makes use of flexible action patterns that do not necessarily require conscious control of one's actions. At the *routine or sensorimotor level* of regulation, automatic movement sequences are performed without conscious attention. As a consequence, action programs that are regulated at the routine or sensorimotor level are very hard to consciously modify when necessary (Frese and Zapf 1994).

For example, if an employee is assigned a new goal to create some sort of report, she or he will first have to make decisions about how to break this task into subtasks (e.g., research up-to-date numbers, process them, write the report) and plan which tools she/he will use (e.g., whether to use a spreadsheet application or a statistical application to process the numbers). Throughout the execution of these plans, the employee will have to monitor her/his progress and adapt her/his plans if necessary. These forms of regulations would most likely take place at the intellectual level, requiring high levels of cognitive processing. During the execution of her or his plans, the employee might, however, also encounter situations that allow for regulation at the knowledge-based and/or the routine or sensorimotor levels. While preparing the data, the employee might realize that some data is only available in a specific format and has to be converted before it can be processed. If the employee has encountered such a situation before and already knows a way to deal with it, regulation will take place on the knowledge-based level, requiring fewer cognitive resources. Similarly, the employee might discover that some of the data is only available in printed form and thus has to be entered into the computer manually. If the employee already has a lot of experience with data entry, regulation can proceed on the routine or sensorimotor level, requiring very few cognitive resources.

Action regulation theory not only elegantly describes the processes involved in the regulation of work tasks (translating tasks into goals, generating action plans, monitoring action, and processing feedback) but also provides an excellent explanation of how work stressors lead to work strain through increased taxation on employees' mental resources for the regulation of action processes.

Based on action regulation theory (e.g., Zapf 2002), task characteristics can be divided into three groups: regulation requirements, regulation possibilities, and regulation problems. Regulation requirements refer to task characteristics like task variety and task complexity. Highly complex tasks require more regulation at the intellectual level than tasks that are less complex, highly routinized, or have been the subject of extensive training. Regulation possibilities include various aspects of how employees can exert control over the work situation like the order and timing in which operations are performed. Finally, regulation problems subsume various forms of work stressors that interfere with and disturb action regulation processes. Regulation problems include, for example, interruptions (because they obstruct action regulation), qualitative overload (because it creates uncertainty in regulation processes), as well as time pressure and other demands that overtax employees' capacity for action regulation (Frese and Zapf 1994). Well-designed jobs are not only characterized by low levels of work stressors that pose regulation problems and lead to health impairment, but also involve adequate levels of regulation requirements and regulation possibilities that provide employees with opportunities for learning and personality development (Glaser et al. 2015).

Building on these theoretical assumptions of how work stressors are linked to indicators of job strain, Diestel and Schmidt (2012) proposed and demonstrated that regulation problems (high levels of qualitative and quantitative workload) trigger resource-depleting self-regulatory processes. By integrating action regulation theory with the self-control strength model (e.g., Baumeister et al. 2007), Diestel and Schmidt (2012) showed that coping with high levels of qualitative and quantitative workload requires employees to inhibit impulsive response tendencies, ignore task-irrelevant stimuli, and overcome motivational deficits to complete unattractive tasks. These cognitive processes, which are a requirement for action regulation, have previously been shown to draw on and deplete limited cognitive resources for self-regulation (Hagger et al. 2010).

Further evidence that regulation problems do indeed express their adverse effects on employee well-being by overtaxing employees' resources for action regulation comes from a recent diary study showing that not only high workload or time pressure but also emotional dissonance triggers resource-depleting self-regulatory processes that require increased self-control effort (Prem et al. 2016). This shows that the hypothesized adverse effects also extend to regulation problems other than high levels of workload. However, in our diary study, my colleagues and I also showed that regulation problems trigger anxiety and that the link between regulation problems and increased self-control effort can in part be explained via these increased levels of anxiety in employees faced with regulatory problems. This indicates that it is important to also consider the emotions triggered in cognitive appraisal processes in order to get a more comprehensive picture of the immediate within-person processes that take place during the workday.

7.1.2 Cognitive Appraisal Processes: Appraisals and Emotions

In contrast to action regulation theory, which has a strong focus on the mental processes underlying goal-oriented work behavior but leaves the individual's subjective perception of the work situation deliberately aside (cf. Frese and Zapf 1994), other theories highlight the role of subjective evaluations in stress processes because of their impact on emotions and coping behavior. Most of these theories draw on the transactional model of stress (Lazarus and Folkman 1984) to explain the role of cognitive appraisal processes in work settings.

The transactional model of stress (Lazarus and Folkman 1984) describes coping behavior as a function of cognitive appraisal processes. In what is known as *primary appraisal*, the individual evaluates whether there is something at stake in the situation ("Am I in trouble or being benefited, now or in the future, and in what way?", Lazarus and Folkman 1984: 31). A situation can be appraised as *irrelevant* if it does not affect the individual's well-being, as *benign-positive* if it preserves or enhances the individual's well-being, or as *stressful* if it has the potential to negatively affect the individual's well-being. If the situation is appraised as stressful, one can distinguish the stress appraisal further into situations where *harm or loss* has already occurred and situations that pose a *threat* and/or *challenge* for the future. Whether the situation is perceived as having only negative effects in the future (threat appraisal) or whether it also has the potential for gain and growth (challenge appraisal) will affect the individual's emotions. In the case of threat appraisals, people are more likely to experience negative emotions such as fear, anxiety, and anger, whereas in the case of challenge appraisal, positive emotions like eagerness, excitement, and exhilaration will dominate.

During *secondary appraisal*, the individual will evaluate her or his coping options ("What if anything can be done about it?", Lazarus and Folkman 1984: 31). In this cognitive process, the individual not only evaluates which coping options she/he has available but also her/his likelihood of being able to execute the coping behavior and the likelihood that the coping behavior leads to the desired outcome. Although the terms primary appraisal and secondary appraisal seem to suggest a chronological order, the two types of cognitive appraisal are presumed to take place in parallel and influence one another. For example, if the available coping options are evaluated as presumably ineffective against an expected harm or loss, the individual will appraise the situation as a threat, whereas if one considers the situation to be controllable, she or he is also more likely to appraise it as a challenge. The cognitive appraisal of a situation will not only elicit certain emotions but also influence coping behavior. *Problem-focused coping* that aims to directly attack the stressor is more likely when the individual perceives control over the situation, whereas other situations, especially those in which the individual perceives negative emotions, will trigger *emotion-focused coping* behavior that aims to alter the interpretation of the stressor (see also Lazarus 1995).

The idea that individuals' cognitive appraisal of a situation affects how they deal with stressful situations has also influenced work psychology. In the challenge stressors-hindrance stressors literature (Cavanough et al. 2000; LePine et al. 2005), scholars distinguish between *challenge stressors* that have the potential to motivate employees and *hindrance stressors* that lack equivalent positive effects. Although both types of stressors require effort and thus draw on and deplete employees' resources, challenge stressors like time pressure and responsibility are thought to also have positive effects that stem from feelings of accomplishment and growth when these stressors are met successfully. In contrast, overcoming hindrance stressors like role ambiguity and red tape does not typically provide such positive experiences. Consequently, hindrance stressors do not promote motivation and are only related to indicators of job strain (e.g., LePine et al. 2005; Crawford et al. 2010; Van den Broeck et al. 2010).

In recent years, research based on the challenge stressors-hindrance stressors framework has not only classified stressors as challenge or hindrance stressors based on their effects but has also begun to measure cognitive appraisal directly (Webster et al. 2011). Diary studies have shown that challenge stressors are indeed more strongly related to *challenge appraisal* than hindrance appraisal and that hindrance stressors are only related to *hindrance appraisal,* not challenge appraisal

(Searle and Auton 2015). Furthermore, it has been shown that challenge stressors are linked to activated positive affect via challenge appraisal and that hindrance stressors are linked to anger via hindrance appraisal (Searle and Auton 2015). Other diary studies have demonstrated that not only job demands like time pressure but also job resources like job control or positive meaning have positive effects on work outcomes like creativity, proactive work behavior, and learning at work via challenge appraisal (Ohly and Fritz 2010; Prem et al. 2017). Finally, it has recently been shown that it makes sense to not only measure challenge appraisal and hindrance appraisal but also to distinguish between hindrance appraisals and *threat appraisals*. Tuckey et al. (2015) showed in a diary study that stressors that obstruct accomplishment promote fatigue via hindrance appraisal, whereas stressors related to harm to the self or personal loss lead to anger and anxiety via threat appraisals. Overall, empirical evidence suggests that cognitive appraisal processes are useful in explaining within-person relationships between work characteristics, emotions and performance-related outcomes.

7.1.3 Motivational Processes: Basic Need Satisfaction and Work Motivation Quality

An overview of the within-person processes that underlie goal-oriented work behavior would not be complete without also considering motivational processes. To get a more comprehensive picture of day-level within-person processes and complement action regulation theory, which focuses on how work tasks are mentally regulated, and the transactional stress model, which explains how employees' appraisal of their work situation impacts their emotions and coping strategies, I will draw on self-determination theory (e.g., Ryan and Deci 2000) to describe the determinants of different kinds of motivation at work.

Self-determination theory (e.g., Ryan and Deci 2000; see also Gagné and Deci 2005) views motivation quality on a continuum from amotivation to various forms of extrinsic motivation and finally to intrinsic motivation. In contrast to *intrinsic motivation*, where behavior is driven by interest in and enjoyment of the task itself, extrinsically motivated behavior is pursued for different reasons. In the case of *extrinsic motivation,* individuals may regulate their behavior simply to attain external rewards or to avoid external punishments (external regulation). In such cases, the individual will only perform actions if she or he expects them to be instrumental in obtaining desired rewards and/or avoiding undesired consequences. For example, an employee might only work if she or he is being monitored by her/his supervisor. However, extrinsic motivation can also be internalized to various degrees. In the case of introjected regulation, employees typically perform their work tasks because they feel pressure to attain set goals to maintain their self-esteem. In contrast, in the case of identified regulation, employees perform a behavior because it is in line with their personal goals and identities. Finally, the most internalized form of external motivation is integrated regulation, which occurs when employees see their behavior as an integral part of themselves (Gagné and Deci 2005).

Although identified regulation and integrated regulation are still forms of external motivation (because they are pursued to attain personal goals), they also share some similarities with intrinsic motivation. As a consequence, identified regulation and integrated regulation are often combined with intrinsic motivation to form a composite known as *autonomous motivation*, which is then distinguished from external regulation and introjected regulation, which are combined to a composite known as *controlled motivation* (Gagné and Deci 2005).

The degree to which a given behavior is internalized is thought to be influenced by aspects of the social context and, in particular, by the satisfaction of basic psychological needs it provides (Ryan and Deci 2000). Self-determination theory argues that the satisfaction of needs for *relatedness*, *competence*, and *autonomy* acts as a nutriment for autonomous motivation (e.g., Gagné and Deci 2005). Internalization of externally motivated behaviors is fostered when an individual feels a sense of belonging and connectedness with others (relatedness), experiences efficacy with regard to the behavior (competence), and has a sense of choice and freedom (autonomy). In work settings, these factors might not only be influenced by aspects of the job content and context, but also by a work climate that can be described as *autonomy supportive*. Managers can foster a climate supportive of autonomy by giving their employees opportunities to take initiative and make choices at work, by providing them with non-controlling informal feedback and by recognizing their needs and feelings (Gagné and Deci 2005; see also Van den Broeck et al. 2008). However, if basic psychological needs are frustrated rather than satisfied, this will be detrimental to the internalization of behaviors and foster controlled motivation instead of autonomous motivation.

It has been assumed that autonomous motivation should have effects on both employees' well-being and their performance (Gagné and Deci 2005). Empirical studies have confirmed that need satisfaction partially mediates between job demands and job resources and indicators of employee well-being like vigor and exhaustion (Van den Broeck et al. 2008). Further, it has been shown that job demands increase controlled motivation via need frustration and that job resources decrease controlled motivation by lowering need frustration and increase autonomous motivation via need satisfaction (Trépanier et al. 2015). Controlled motivation has been linked to more psychological distress, more psychosomatic complaints, lower work engagement, and lower job performance, whereas autonomous motivation has been shown to be associated with higher work engagement and better job performance (Trépanier et al. 2015). Need frustration has also been used to explain the adverse effects of job insecurity on emotional exhaustion and vigor (Vander Elst et al. 2012) as well as counterproductive work behavior from both an organizational and interpersonal perspective (Van den Broeck et al. 2014). Regarding within-person processes, diary studies have shown that within-person processes serially link need satisfaction to vigor and fatigue via increased intrinsic work motivation and subsequently reduced self-control effort (Van Hooff and

Geurts 2015). Overall, empirical evidence seems to support the assumption that basic need satisfaction and/or basic need frustration affect employees' well-being and performance via autonomous motivation and/or controlled motivation. Moreover, results from diary studies suggest that these processes are also linked to self-control effort.

7.1.4 The Interplay of Within-Person Processes in Daily Working Life

Action regulation processes, cognitive appraisal processes, and motivational processes unfold simultaneously and influence one another in daily working life. In the following section, I will try to integrate these processes in a framework model to draw a more complete picture of the interplay of the within-person processes that impact employees' well-being and performance. Integrating these three different processes requires me to reduce some complexity, and thus I will sacrifice some accuracy and richness of the theories that I draw upon to obtain a reasonably simple framework model.

As depicted in Fig. 7.1, I chose to classify work characteristics into hindrance demands, challenge demands, and job resources based on the challenge stressor-hindrance stressor literature (cf. Crawford et al. 2010).[1] Based on action regulation theory (e.g., Frese and Zapf 1994), I assume that both hindrance demands and challenge demands will trigger action regulation processes because they ought to pose regulation requirements and/or regulation problems. Job resources will act as regulation possibilities and thus facilitate action regulation processes. Action regulation processes, in turn, are thought to require self-control effort (e.g., Diestel and Schmidt 2012) and thus draw on and deplete employees' mental resources. As a consequence, self-control effort will negatively affect both employees' well-being and their performance. However, action regulation processes might also foster personality development and thus lead to learning, which could have positive effects on employees' well-being and performance in the long run.

From the challenge stressors-hindrance stressors literature (e.g., Searle and Auton 2015; Tuckey et al. 2015), it can be assumed that hindrance demands promote hindrance and/or threat appraisals that in turn increase negative emotions

[1] I chose this classification of work characteristics because of its popularity in international work and organizational psychology literature. However, it should be mentioned that this classification does not fully overlap with action regulation theory. Although hindrance demands (e.g., red tape) are usually regarded as regulation problems, job resources (e.g., job autonomy) are considered regulation possibilities, and challenge demands can often be described as regulation requirements (e.g., responsibility), there are some exceptions to this rule of thumb. Most importantly, time pressure is regarded not as a regulation requirement but a regulation problem in action regulation theory.

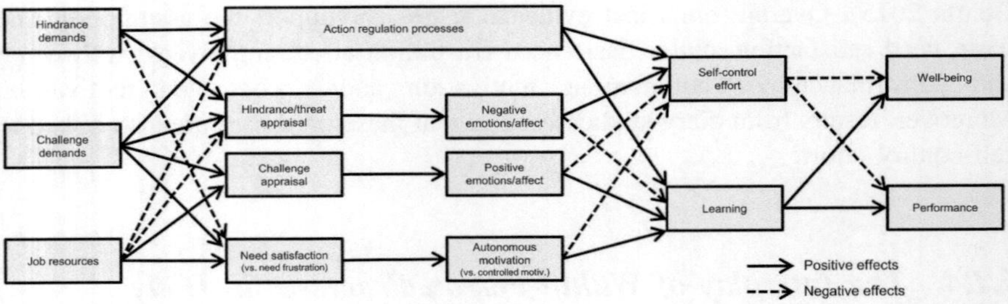

Fig. 7.1 Framework model that integrates action regulation processes, cognitive appraisal processes, and motivational processes

and/or negative affect. Because negative emotions entrain action tendencies that lead to withdrawal and/or aggressive behavior, employees will have to invest self-control effort to overcome these action tendencies (e.g., Mackey and Perrewé 2014). Challenge demands should promote both challenge appraisals and hindrance appraisals (e.g., Webster et al. 2011; Searle and Auton 2015; Prem et al. 2017). Via their effects on challenge appraisal, challenge demands should increase positive affect, which might provide employees not only with energy that can be invested in self-control effort but also broaden their thought-action repertoire (Fredrickson 2001), promoting creativity, development, and learning at work (Prem et al. 2017). Because job resources help employees cope with their hindrance and challenge demands, job resources should lead to more challenge appraisals and fewer hindrance appraisals (e.g., Ohly and Fritz 2010; Paškvan et al. 2016; Prem et al. 2017), thus reducing self-control effort and boosting learning at work.

Finally, according to self-determination theory (e.g., Ryan and Deci 2000; Gagné and Deci 2005), hindrance demands should frustrate basic psychological needs and thus have negative effects on need satisfaction, while challenge demands and job resources should provide opportunities for need satisfaction (cf. Van den Broeck et al. 2010; Albrecht 2015). Need satisfaction should in turn lead to more autonomous motivation, which should decrease self-control effort (cf. Van Hooff and Geurts 2015) and promote employees' learning at work (Spreitzer and Porath 2014).

In my framework model (cf. Fig. 7.1), self-control effort and learning take central roles in combining the effects of action regulation processes, cognitive appraisal processes, and motivational processes. It is assumed that adverse effects on employees' health and performance will mainly be explained via self-control effort and beneficial effects on employees' health and performance will mainly be explained via learning at work.

By containing both adverse as well as beneficial indirect effects on work outcomes, the framework model shares some similarity with other models of job stress that differentiate between (adverse) strain-related processes and (beneficial) motivational processes (e.g., Bakker and Demerouti 2007; Glaser et al. 2015; LePine et al. 2005). Because the effects via self-control effort and learning at work often antagonize each other, the overall effect of work characteristics on work outcomes will be ambivalent and the result in the long run will depend on whether the adverse effects outweigh the beneficial effects or vice versa.

It should be mentioned that I do not assume that self-control effort and learning at work will be able to completely explain the effects of hindrance demands, challenge demands, and job resources on employees' well-being and performance. Nevertheless, self-control effort and learning at work have a prominent role in the framework model because they link within-person processes of action regulation, cognitive appraisal, and motivation.

7.2 The Effects of Changing Work Environments on Within-Person Processes

In the following section, I will reflect on how changes in modern work environments might affect the previously described within-person processes of action regulation, cognitive appraisal, and motivation in daily working life. At first glance, it may seem counterintuitive that changes in work environments (that oftentimes unfold over relatively long periods of time) can have an impact on processes that unfold from day to day (within rather short periods of time). However, some earlier theories of work stress have already made assumptions about how chronic work characteristics might affect within-person stress processes and vice versa. For example, the allostatic load model (McEwen 2007) explains how biological processes of short-term stress responses lead to more chronic changes that alter the homeostasis in endocrinal and neuronal mechanisms and how such adaptation processes can affect future short-term stress responses. Largely on the basis of McEwen's model, Ilies and colleagues called for researchers to investigate the interaction of short-term and chronic effects in naturalistic settings and to develop other comprehensive theoretical models integrating the perspectives of short-term and chronic effects (Ilies et al. 2015, 2016).

To explain how changes in work environments affect short-term within-person processes, I will add another layer of complexity and differentiate between (a) long-term changes in work environments that affect both (b) chronic working conditions and (c) day-level within-person processes. Changing work environments should alter chronic work characteristics over time. As a consequence, long-term changes in the work environment may affect day-level within-person processes via their effects on chronic work characteristics. For example, work intensification may lead to more chronic workload over time, which may influence how employees cope with daily time pressure. However, it is also plausible that not all effects of changing work environments on day-level within-person processes are mediated through chronic work characteristics. For example, work intensification might not only affect how employees cope with daily time pressure because it leads to higher chronic workload, but also more directly, because employees experiencing work intensification might develop a more negative attitude towards workload and time pressure as a consequence of work intensification. In other words, changes in the work environment may not only affect day-level within-person processes because they alter the quantitative level of chronic work characteristics but also because they might alter the appraisal of work characteristics.

7.2.1 Effects of Changing Work Environments on Action Regulation Processes

Changes in modern work settings may affect the action regulation process in various ways. First, if chronic work characteristics change, regulation demands may increase and, if they exceed certain levels, also pose additional regulation problems. For example, work intensification increases chronic workload and consequently also daily time pressure. If employees have to work faster than usual, their capacity for action regulation might be overtaxed, meaning that additional self-control effort is required. Of course, it should be mentioned that changes in work environments do not necessarily have only adverse effects. For example, redesigning traditional office spaces into activity-based flexible office spaces might also reduce noise and interruptions and thus reduce regulation problems.

Second, frequent changes in the way work is organized and/or the technical equipment used to carry out work will require employees to update action plans or generate new ones more often. This will increase action regulation processes that unfold at the intellectual level. As a consequence, employees will be less able to rely on regulation processes at the knowledge-based level and/or the routine or sensorimotor level. Because they are less able to rely on routines, additional self-control effort will again be required from employees. However, increased regulation at the intellectual level might also provide opportunities for development and learning at work.

Third, some changes in the work environment will require employees to not only regulate goal-oriented work behavior, but also to get involved in other self-regulation processes. For example, as work becomes more flexible with respect to where and when employees engage in their work, employees will have to plan how they organize their work tasks during working hours but also the times at which they work and when they perform chores or spend time with their family, etc. Although such forms of flexibility provide employees with opportunities to adapt their work schedule to non-work needs, the self-regulation processes involved in planning not only one's actions at work but also when and where to work require additional mental resources. Similarly, if career paths become less stable and employees have to take care of their own development and future employment opportunities, employees will have to invest additional resources not only in making plans for their career but also in pursuing these plans.

Finally, some work characteristics that are usually regarded as resources for regulation might acquire the quality of regulation requirements. For example, having control over the order in which one performs individual work tasks and the means one uses to do so are usually regarded as resources that ease action regulation processes. However, if such work characteristics are combined with high performance goals and no clear structure of how to obtain these goals, they may no longer provide employees with freedom in their work. Instead, employees might be left only with increased qualitative demands that require additional self-control effort in goal-oriented work behavior.

7.2.2 *Effects of Changing Work Environments on Cognitive Appraisal Processes*

Changes in modern work settings may also affect cognitive appraisal processes in multiple ways. First, changes in the extent of hindrance demands should have effects on the extent to which employees appraise their work situations as hindrances and/or threats. If organizations reduce bureaucracy, employees might experience less red tape, which should also reduce hindrance appraisals as well as experiences of negative emotions and negative affect. At the same time, if competences are less clearly defined, hindrance stressors like role ambiguity might increase, which could lead employees to perceive their work situation as more hindering. Similarly, increases in job insecurity might boost threat appraisals in employees, accompanied by feelings of anger and/or anxiety.

Second, changes in the extent of challenge demands should have effects on challenge appraisals. Increases in time pressure and responsibility should theoretically facilitate challenge appraisals as well as positive emotions and positive affect. As a consequence, increasing challenge demands might provide employees with opportunities for learning and growth at work. However, it should be highlighted that challenge demands can also be appraised as hindrances. In particular, if time pressure reaches a level where employees no longer expect to be able to attain their goals, they should be more likely to appraise their work situations as a hindrance and/or threat, which goes along with negative emotions and negative affect.

Finally, because job resources play an important role in secondary appraisal processes when employees evaluate their coping options, the availability of job resources should also impact challenge appraisals, hindrance appraisals, and/or threat appraisals. In particular, job resources should boost challenge appraisals and extenuate hindrance and/or threat appraisals.

7.2.3 *Effects of Changing Work Environments on Motivational Processes*

In addition to their potential effects on action regulation processes and cognitive appraisal processes, changes in modern work settings should also impact motivational processes. Changes in working conditions may affect the extent to which basic needs are satisfied and/or frustrated, which in turn affect the quality of motivation.

On the one hand, the frustration of basic psychological needs should lead to more controlled motivation, which goes along with increased self-control effort and less learning at work. For example, when facing increasing levels of workload due to work intensification, employees may no longer be able to attain their work goals and consequently cease to experience feelings of competence. As a consequence, they will experience less autonomous motivation and more controlled motivation. This ultimately means that these employees will have to invest additional self-control effort to force themselves to work.

On the other hand, modern work settings may also aid in the satisfaction of basic psychological needs. This promotes more autonomous motivation, which should reduce self-control effort and provide a climate for learning at work. For example, flexible work times that give employees true freedom over when they work could help satisfy employees' need for autonomy and consequently lead to autonomous motivation. Similarly, if a worker decides to switch to a job or profession that is more in line with their personal goals and identities, they may also be more autonomously motivated at work.

7.3 Ambivalent Consequences of Changing Work Environments in Daily Working Life

As we have seen in the previous chapters, societal, economic, and technological changes affect modern work settings. This may result in both adverse and beneficial effects on work outcomes like employee well-being and performance in daily working life.

Changes that lead to regulation problems, promote hindrance or threat appraisals, and/or frustrate basic psychological needs will likely require employees to invest additional self-control effort at work. Because the investment of self-control effort draws on and depletes employees' limited resources for self-regulation, this will have adverse effects on employees' job strain. Longitudinal studies have shown that demands to invest self-control effort at work are related to indicators of burnout like emotional exhaustion and depersonalization (Diestel and Schmidt 2012). Further, diary studies have also linked requirements to invest self-control effort to increased need for recovery and ego depletion as well as decreased work engagement (Rivkin et al. 2015; Prem et al. 2016). It can be assumed that if employees are not able to sufficiently recover the resources they invest in additional self-control effort, it will not only have effects on well-being but also affect their work performance. For example, if employees' self-regulatory resources are depleted, they may be more likely to procrastinate on their work tasks (Kühnel et al. 2016) and may also lack the energy to be proactive (Parker et al. 2010).

Changes that foster regulation requirements, promote challenge appraisals, and/or satisfy basic psychological needs will likely also provide employees with opportunities for learning and growth. Research on employees' thriving at work has shown that employees who experience both vitality as well as a sense of learning and personal growth at work not only score lower on burnout indicators but are also more likely to be satisfied with their job and committed to their organizations (Porath et al. 2012). Furthermore, thriving employees are also more proactive at work, show more initiative in their career development, and attain better job performance (Porath et al. 2012). However, it should be highlighted that to experience thriving at work, it is not sufficient to experience just a sense of learning and personal growth at work. Instead, thriving individuals also have to experience vitality, which can be described as a sense of aliveness (cf. Spreitzer et al. 2005). Thus, employees will not be able to thrive on learning opportunities if their mental resources are drained by the self-control effort they have to put into their work.

Although some changes, like work intensification, may have mainly adverse consequences on work outcomes, other changes, like those related to flexible working, may also hold the potential for beneficial effects. Overall, one can assume that many societal, economic, and technological changes may have ambivalent consequences on within-person processes of action regulation, cognitive appraisal, and motivation as well as on work outcomes that result from these processes.

7.4 Conclusion

In this chapter I gave an overview of various within-person processes that link work characteristics to work outcomes in daily working life. Drawing on action regulation theory (e.g., Frese and Zapf 1994; Hacker 2003), the transactional stress model (Lazarus and Folkman 1984; Searle and Auton 2015), as well as self-determination theory (Ryan and Deci 2000; Gangné and Deci 2005), I tried to explain how changes in modern work settings affect within-person processes in daily working life.

Because societal, economic, and technological changes may have both adverse and beneficial effects on employees' well-being and performance, one might conclude that is not yet clear whether these changes will be more of a blessing or a curse for employees in the future. Although this might be true in some cases, like flexible working arrangements (Gerdenitsch et al. 2015), there is also some evidence suggesting that a few changes are clearly negative. For example, it has been shown that workers appraise work intensification as a hindrance (Paškvan et al. 2016). To predict how changes in work settings affect work outcomes, it may thus make sense to consider their effects on within-person processes in daily working life.

I recommend that future research not only tries to further integrate theories about various within-person processes but also consider how these processes might interact on the short-, medium-, and long-term levels. In my opinion, it is important to empirically analyze how changes in modern work settings affect chronic work characteristics and how these changes affect within-person processes in daily working life (see also Ilies et al. 2015, 2016).

The changes in today's world of work pose challenges for companies. In order to stay competitive, companies have to adapt quickly to societal, economic, and technological changes in the work environment. However, it is also becoming increasingly important for companies to maintain a thriving workforce that is energized, grows and develops to be sustainable (Spreitzer et al. 2012). Thus, it is necessary to actively create working conditions that enable employees to stay both healthy and productive. Creating work climates supportive of autonomy and ensuring that employees have sufficient opportunities to recharge their energy and stay vital could be key to fostering the conditions that allow both employees and their companies to thrive.

Finally, I would like to remind practitioners it is upon them to actively design tomorrow's work environments. Although practitioners may consider themselves constrained in how they can react to the effects of societal, economic, and technological changes on work settings, they should focus on how to buffer the adverse effects of such changes through work redesign while also keeping an eye on how to utilize potential beneficial effects. Employees should be given opportunities to craft their jobs and/or their careers to match their needs. These efforts should be backed by legislation that emphasizes the importance of healthy work environments. In the end, well-designed jobs will not only benefit employees themselves but also the companies they work for and society as a whole.

References

Albrecht, S. L. (2015). Challenge demands, hindrance demands, and psychological need satisfaction. *Journal of Personnel Psychology, 14,* 70–79.

Bakker, A. B., & Demerouti, E. (2007). The job demands-resources model: State of the art. *Journal of Managerial Psychology, 22,* 309–328.

Baumeister, R. F., Vohs, K. D., & Tice, D. M. (2007). The strength model of self-control. *Current Directions in Psychological Science, 16,* 351–355.

Bolger, N., Davis, A., & Rafaeli, E. (2003). Diary methods: Capturing life as it is lived. *Annual Review of Psychology, 54,* 579–616.

Cascio, W. F. (1995). Whither industrial and organizational psychology in a changing world of work? *American Psychologist, 50,* 928–939.

Cavanaugh, M. A., Boswell, W. R., Roehling, M. V., & Boudreau, J. W. (2000). An empirical examination of self-reported work stress among US managers. *Journal of Applied Psychology, 85,* 65–74.

Crawford, E. R., LePine, J. A., & Rich, B. L. (2010). Linking job demands and resources to employee engagement and burnout: A theoretical extension and meta-analytic test. *Journal of Applied Psychology, 95,* 834–848.
Diestel, S., & Schmidt, K.-H. (2012). Lagged mediator effects of self-control demands on psychological strain and absenteeism. *Journal of Occupational and Organizational Psychology, 85,* 556–578.
Fisher, C. D., & To, M. L. (2012). Using experience sampling methodology in organizational behavior. *Journal of Organizational Behavior, 33,* 865–877.
Fredrickson, B. L. (2001). The role of positive emotions in positive psychology: The broaden-and-build theory of positive emotions. *American Psychologist, 56,* 218−226.
Frese, M., & Zapf, D. (1994). Action as the core of work psychology: A German approach. In H. C. Triandis, M. D. Dunnette, & L. M. Hough (Eds.), *Handbook of industrial and organizational psychology* (Vol. 4, pp. 271–340). Palo Alto, CA: Consulting Psychologists Press.
Gagné, M., & Deci, E. L. (2005). Self-determination theory and work motivation. *Journal of Organizational Behavior, 26,* 331–362.
Gerdenitsch, C., Kubicek, B., & Korunka, C. (2015). Control in flexible working arrangements: When freedom becomes duty. *Journal of Personnel Psychology, 14,* 61–69.
Glaser, J., Seubert, C., Hornung, S., & Herbig, B. (2015). The impact of learning demands, work-related resources, and job stressors on creative performance and health. *Journal of Personnel Psychology, 14,* 37–48.
Hacker, W. (2003). Action regulation theory: A practical tool for the design of modern work processes? *European Journal of Work and Organizational Psychology, 12,* 105–130.
Hackman, J. R. (1970). Tasks and task performance in research on stress. In J. E. McGrath (Ed.), *Social and psychological factors in stress* (pp. 202–237). New York, NY: Holt, Rinehart & Winston.
Hagger, M. S., Wood, C., Stiff, C., & Chatzisarantis, N. L. (2010). Ego depletion and the strength model of self-control: A meta-analysis. *Psychological Bulletin, 136,* 495–525.
Ilies, R., Aw, S. S., & Lim, V. K. (2016). A naturalistic multilevel framework for studying transient and chronic effects of psychosocial work stressors on employee health and well-being. *Applied Psychology, 65,* 223–258.
Ilies, R., Aw, S. S., & Pluut, H. (2015). Intraindividual models of employee well-being: What have we learned and where do we go from here? *European Journal of Work and Organizational Psychology, 24,* 827–838.
Kubicek, B., Paškvan, M., & Korunka, C. (2015). Development and validation of an instrument for assessing job stressors arising from accelerated change. The intensification of job demands Scale (IDS). *European Journal of Work and Organizational Psychology, 24,* 898–913.
Kühnel, J., Bledow, R., & Feuerhahn, N. (2016). When do you procrastinate? Sleep quality and social sleep lag jointly predict self-regulatory failure at work. *Journal of Organizational Behavior, 37,* 983–1002.
Lazarus, R. S. (1995). Psychological stress in the workplace. In R. Crandall & P. L. Perrewé (Eds.), *Occupational stress: A handbook* (pp. 3–19). Washington, DC: Taylor & Francis.
Lazarus, R. S., & Folkman, S. (1984). *Stress, appraisal, and coping*. New York, NY: Springer.
LePine, J. A., Podsakoff, N. P., & LePine, M. A. (2005). A meta-analytic test of the challenge stressor–hindrance stressor framework: An explanation for inconsistent relationships among stressors and performance. *Academy of Management Journal, 48,* 764–775.
Mackey, J. D., & Perrewé, P. L. (2014). The AAA (appraisals, attributions, adaptation) model of job stress: The critical role of self-regulation. *Organizational Psychology Review, 4,* 258–278.
McEwen, B. S. (2007). Physiology and neurobiology of stress and adaptation: Central role of the brain. *Physiological Reviews, 87,* 873–904.
Ohly, S., & Fritz, C. (2010). Work characteristics, challenge appraisal, creativity, and proactive behavior: A multi-level study. *Journal of Organizational Behavior, 31,* 543–565.

Ohly, S., Sonnentag, S., Niessen, C., & Zapf, D. (2010). Diary studies in organizational research: An introduction and some practical recommendations. *Journal of Personnel Psychology, 9,* 79–93.

Parker, S. K., Bindl, U. K., & Strauss, K. (2010). Making things happen: A model of proactive motivation. *Journal of Management, 36,* 827–856.

Paškvan, M., Kubicek, B., Prem, R., & Korunka, C. (2016). Cognitive appraisal of work intensification. *International Journal of Stress Management, 23,* 124–146.

Porath, C., Spreitzer, G., Gibson, C., & Garnett, F. G. (2012). Thriving at work: Toward its measurement, construct validation, and theoretical refinement. *Journal of Organizational Behavior, 33,* 250–275.

Prem, R., Kubicek, B., Diestel, S., & Korunka, C. (2016). Regulatory job stressors and their within-person relationships with ego depletion: The roles of state anxiety, self-control effort, and job autonomy. *Journal of Vocational Behavior, 92,* 22–32.

Prem, R., Ohly, S., Kubicek, B., & Korunka, C. (2017). Thriving on challenge stressors? Exploring time pressure and learning demands as antecedents of thriving at work. *Journal of Organizational Behavior, 38,* 108–123.

Rivkin, W., Diestel, S., & Schmidt, K. H. (2015). Affective commitment as a moderator of the adverse relationships between day-specific self-control demands and psychological well-being. *Journal of Vocational Behavior, 88,* 185–194.

Rosa, H. (2003). Social acceleration: Ethical and political consequences of a desynchronized high–speed society. *Constellations, 10,* 3–33.

Rosa, H. (2013). *Social acceleration: A new theory of modernity*. New York, NY: Columbia University Press.

Ryan, R. M., & Deci, E. L. (2000). Self-determination theory and the facilitation of intrinsic motivation, social development, and well-being. *American Psychologist, 55,* 68–78.

Searle, B. J., & Auton, J. C. (2015). The merits of measuring challenge and hindrance appraisals. *Anxiety, Stress, & Coping, 28,* 121–143.

Sonnentag, S. (2001). Work, recovery activities, and individual well-being: A diary study. *Journal of Occupational Health Psychology, 6,* 196–210.

Sonnentag, S., & Bayer, U. V. (2005). Switching off mentally: Predictors and consequences of psychological detachment from work during off-job time. *Journal of Occupational Health Psychology, 10,* 393–414.

Sonnentag, S., & Fritz, C. (2007). The Recovery Experience Questionnaire: Development and validation of a measure for assessing recuperation and unwinding from work. *Journal of Occupational Health Psychology, 12,* 204–221.

Sonnentag, S., & Fritz, C. (2015). Recovery from job stress: The stressor-detachment model as an integrative framework. *Journal of Organizational Behavior, 36,* S72–S103.

Spreitzer, G. M., & Porath, C. (2014). Self-determination as nutriment for thriving: Building an integrative model of human growth at work. In M. Gagné (Ed.), *Oxford handbook of work engagement, motivation, and self-determination theory* (pp. 245–258). New York, NY: Oxford University Press.

Spreitzer, G., Porath, C. L., & Gibson, C. B. (2012). Toward human sustainability: How to enable more thriving at work. *Organizational Dynamics, 41,* 155–162.

Spreitzer, G., Sutcliffe, K., Dutton, J., Sonenshein, S., & Grant, A. M. (2005). A socially embedded model of thriving at work. *Organization Science, 16,* 537–549.

Trépanier, S. G., Forest, J., Fernet, C., & Austin, S. (2015). On the psychological and motivational processes linking job characteristics to employee functioning: Insights from self-determination theory. *Work & Stress, 29,* 286–305.

Tuckey, M. R., Searle, B. J., Boyd, C. M., Winefield, A. H., & Winefield, H. R. (2015). Hindrances are not threats: Advancing the multidimensionality of work stress. *Journal of Occupational Health Psychology, 20,* 131–147.

Van den Broeck, A., De Cuyper, N., De Witte, H., & Vansteenkiste, M. (2010). Not all job demands are equal: Differentiating job hindrances and job challenges in the job

demands-resources model. *European Journal of Work and Organizational Psychology, 19,* 735–759.
Van den Broeck, A., Vansteenkiste, M., De Witte, H., & Lens, W. (2008). Explaining the relationships between job characteristics, burnout, and engagement: The role of basic psychological need satisfaction. *Work & Stress, 22,* 277–294.
Van den Broeck, A., Sulea, C., Vander Elst, T., Fischmann, G., Iliescu, D., & De Witte, H. (2014). The mediating role of psychological needs in the relation between qualitative job insecurity and counterproductive work behavior. *Career Development International, 19,* 526–547.
Van Hooff, M. L. M., & Geurts, S. A. E. (2015). Need satisfaction and employees' recovery state at work: A daily diary study. *Journal of Occupational Health Psychology, 20,* 377–387.
Vander Elst, T., Van den Broeck, A., De Witte, H., & De Cuyper, N. (2012). The mediating role of frustration of psychological needs in the relationship between job insecurity and work-related well-being. *Work & Stress, 26,* 252–271.
Webster, J. R., Beehr, T. A., & Love, K. (2011). Extending the challenge-hindrance model of occupational stress: The role of appraisal. *Journal of Vocational Behavior, 79,* 505–516.
Zapf, D. (2002). Emotion work and psychological well-being: A review of the literature and some conceptual considerations. *Human Resource Management Review, 12,* 237–268.

Author Biography

Roman Prem is a postdoctoral researcher at the University of Vienna, Austria, where he recently received his Ph.D. in work and organizational psychology. His current research focuses on the role of cognitive appraisal and self-control in occupational stress processes. He is also interested in interactions between stress processes, motivational processes, and recovery processes as well as the antecedents of procrastination and proactive behavior at work.

Author Biography

Chapter 8
Challenges for Job Design

Christian Korunka

8.1 New Demands and New Challenges in a Changing World of Work

As was shown in the previous chapters (see especially Flecker et al., Chap. 2, but also the other chapters) there are many changes and challenges in the global world of work, leading to new demands on employees. Global trade markets, increasing competition in nearly every part of society, acceleration processes, deregulation on different levels and the introduction of new information and communication technologies (ICT) are only some of the most important changes and developments over the past few decades. All these trends and developments have strongly influenced today's world of work. It is important to acknowledge that, in general, most of these developments have at least the potential for positive effects, but at the same time they contain many new challenges or even threats for employees. In many circumstances, the benefits of these changes and developments far outweigh the disadvantages, but – as shown in the previous chapters – there are new challenges and demands which have to be taken seriously from a quality of working life perspective.

8.2 Information and Communication Technologies as an Important Part of the Current World of Work

Information and communication technologies are at the core of the societal acceleration processes leading to new job demands. They are important triggers of change at work, tools for completing work and at the same time outcomes of change

C. Korunka (✉)
Faculty of Psychology, University of Vienna, Universitätsstraße 7, 1010 Vienna, Austria
e-mail: christian.korunka@univie.ac.at

C. Korunka and B. Kubicek (eds.), *Job Demands in a Changing World of Work*,
DOI 10.1007/978-3-319-54678-0_8

processes. The widespread use of ICT serves also as an especially good example for the complex pattern of positive effects, negative effects and challenges in modern workplaces (see Table 8.1). The effects of ICT are extremely wide-ranging. On the one hand, ICT unequivocally provide improvements in many life domains. Numerous positive effects on individuals, their jobs, organizations, life outside work and society are observable. In many jobs, working without ICT is no longer conceivable. We all have easy access to a wide range of information, which supports both our working and private lives. ICT support many work tasks and may help to improve aspects of job design, like autonomy. They enable access to all kinds of information for many people. They offer new opportunities for learning and have many other supportive positive features. Job autonomy may increase because of workers' independent and comprehensive access to all kinds of information (see Table 8.1).

On the other hand, there are many challenges related to the widespread use of ICT. Many of the new demands which are described in the earlier chapters are related to the use of ICT. They may even be an outcome of the increased use of ICT. New information technologies are described as the most important driver of acceleration processes (Rosa 2005). Work intensification – in addition to time pressure – is often particularly observable in jobs demanding comprehensive use of

Table 8.1 Examples of challenges and possible positive and negative outcomes related to ICT (Korunka and Hoonakker 2014)

	Challenges	Potential positive outcomes	Potential negative outcomes
Individual level	Need for ICT related competencies Adequate recovery from work Finding a personal work-nonwork balance	Availability and easy access to a wide range of information Increased independence because of better information	Techno strain Techno addiction Alienation
Job	Work intensification Work interruptions Lack of training Frustration because of technical problems	More and better feedback Many new opportunities to learn	Burnout
Organization	Micromanagement Need for development of trust Permanent adaptation needs	Higher quality service Increased productivity Cost reductions	ICT dependency Complex interdependencies between subsystems of organizations
Work-Family	Adequate telework design	Enhanced productivity More "quality time"	Reduced privacy
Society and specific risk groups	Need for new forms of learning Development of opportunities for people with disabilities	Bridging distances Inclusion of all minority groups	ICT dependency

ICT (Chesley 2014). As a result of rapid technological developments, an inherent characteristic of ICT is the permanent need for system improvements and system updates (software updates, hardware updates, new tools, etc.). This results in an ever-increasing need for workers to acquire new competencies vis-a-vis using new forms of information and communication. Plus, there are additional challenges related to ICT use. Work interruptions – often described as an important work stressor (e.g., Frese and Zapf 1994) – may increase, not only because of hardware disruptions but also due to constant use of email and other forms of electronic communication while at work. ICT also provide new forms of control for the organizations (see Table 8.1).

In some cases, there may even be negative effects observable when using ICT. For instance, techno-strain, as a specific form of job related strain (e.g., Brod 1984; Salanova et al. 2013), or even techno-addiction, somewhat comparable to workaholism (e.g., Schaufeli et al. 2008) may occur. Unnoticed by many users, ICT dependency increases. This may not be a problem as long as ICT systems are functioning, but there can be very strong negative effects (strain, productivity losses, etc.) when system breakdowns occur.

8.3 Good Theoretical Framework Models Dealing with Complexity Are Needed

New demands show a complex pattern of numerous positive effects, challenges and negative effects. Such a constellation requires job and organizational design measures that are able to deal with such complex relations. Framework models that are able to describe these complex patterns are needed. The "balance model of job stress", developed more than 35 years ago, was one of the first theoretical frameworks which not only explicitly included technology as a model element (Smith and Sainfort 1989; Carayon and Smith 2014) but is also useful as a general approach to job design when dealing with complex environments and demands with an inherent potential for positive effects, challenges and negative outcomes. The balance model is based on systems theory. It aims to shed light on the complex interactions between certain elements of work design. It postulates that there are certain demands in the work environment that have the potential to lead to a stress/well-being response. By providing a "balance" in job design between demands, the negative effects of stressors on the work environment can be reduced, while the positive effects can be preserved or even enhanced. Another core concept of the model is the idea that positive aspects of jobs and the job environment can compensate for negative effects of the job that cannot be avoided or mitigated. The balance model describes interactions between five components: technology and tools, the organization, the tasks, the environment, and the person (see Fig. 8.1).

Each of the five system elements has certain inherent characteristics that may have positive effects, reduce stress or even increase stress. For instance, elements of

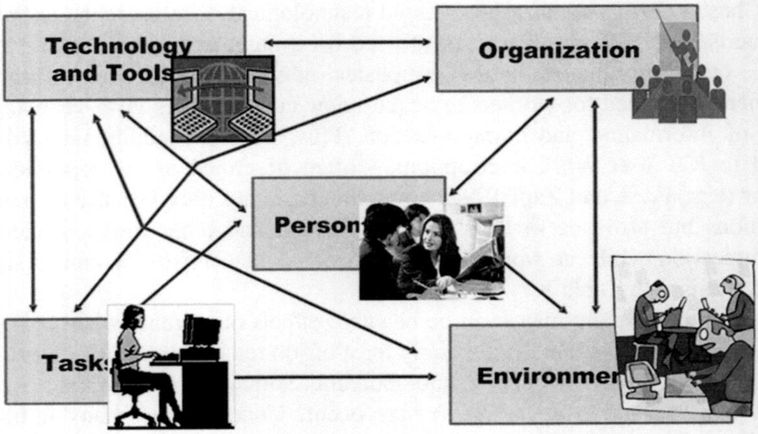

Fig. 8.1 The "balance model" of job stress. *Source* Carayon and Smith (2014), p. 111

the environment, e.g., noise, a problematic light situation or limited space may induce stress. On the other hand, the environment can be designed in such a way that negative effects will be reduced, e.g., by designing the work environment in alignment with ergonomic guidelines or offering need-oriented office spaces in a flexible office concept (e.g., Leather et al. 1998). An essential aspect of the model is to understand the interactions between the five components in order to reach balance in the work system by reducing or compensating for deficits in the system and maintaining/supporting positive aspects at the same time.

Information and communication technologies are one component of the work system. They strongly influence the whole work system and are part of many interactions with the other components of the work system. Thus, "balancing" the effects of IT has many facets. Just a few examples: At many workplaces, ICT may be used flexibly from many locations, including home (e.g., email access). Thus, there is a strong need to balance personal needs in these different work spheres. These needs may be different for persons with families and small children and members of the elderly workforce. To deal with the rapid pace of evolving technologies, one needs to develop and apply skills to keep up with evolving technologies. If IT-related problems occur, one needs to have access to IT support and/or have the knowledge and opportunity to fix the problem. When working with ICT and especially with video screens, one needs a certain number of breaks to balance these demands (Carayon and Smith 2014).

Thus, the "balance model" offers a useful general framework for developing job and organizational design measures in the context of new demands in a changing world of work. Such a framework is a good starting point since it expands our understanding of job design measures by emphasizing the complex interactions between design elements. On the other hand, the balance model is a very general approach to job design. A number of employee needs are not taken into consideration.

8.4 Psychological Needs as a Guideline for Job Design

To deal with the complex demands of the current world of work, considering individual needs is becoming more and more important. ICT can again serve as a good example: Email is a necessary communication tool in many organizations. When using email, differences in space and time become less important. Using smart phones, emails can be sent from anywhere (also from outside the office) at any time (even during night hours). In many organizations, this has led to new challenges. Employees feel pressure to respond promptly when they receive work-related emails outside their usual working times (e.g., Future Work Centre 2016). Some organizations have responded to these new developments by strictly blocking their email systems during non-work hours. This, however, is perceived by other employees as a limitation of their freedom to receive information and work according to their personal schedules.

This well-known case of new demands and approaches for dealing with these demands (e.g., Barley et al. 2011) clearly shows that there are not always adequate strategies and responses available to deal with the needs of all the diverse groups of employees within an organization. This clearly shows that job design measures are becoming even more complex in a changing world of work. We need to deal not only with a complex pattern of positive effects, challenges and negative effects in complex interactions between different design elements, but also with the specific needs of different groups of employees. As a consequence, job design in a given organization should not only take into account the specific needs and demands of the organization, but also of different groups of employees ("differential work design"; e.g., Ulich 1978).

In some cases (as in the example of email use outside working hours), the needs of different groups of employees may differ. Good job design should take such specific needs into consideration. As a first step, employees' specific needs should be found out via interviews or focus groups. These needs should then be taken into consideration in developing a job design solution. Coming back to the aforementioned email example, this would result for instance in flexible rules of email use for different groups of workers.

While there are some circumstances where employees' needs may differ, the satisfaction of general needs may be a good indicator when dealing with the challenges of job demands in the current world of work. Self-determination theory (e.g., Ryan and Deci 2000) is currently the most widely used and widely accepted comprehensive motivational theory dealing not only with general psychological needs, but also with needs at work. Deci and Ryan (2000, p. 229) describe needs as "innate psychological nutriments that are essential for ongoing psychological growth, integrity and well-being." In the original theory, three basic needs are described. “Need for autonomy” is defined as people’s desire to experience personal ownership of their behaviors and act with a sense of volition (Deci and Ryan 2000). It is important to understand that a sense of volition is not only the result of having the opportunity to make personal choices, but also when an externally induced request is fully endorsed, i.e. a meaningful rationale for executing the request is given and one’s personal feelings and needs are acknowledged at the same time. “Need for relatedness” is defined as the human striving for close and intimate relationships and the desire to achieve a state of belonging. Employees who are able to express work-related problems and who really feel part of their work group or team are likely to have their need for relatedness at work fulfilled, in contrast to employees who feel alone at work. Finally, “need for competence” represents people’s desire to feel capable of mastering the environment, achieving the desired outcomes, and managing various challenges (Van den Broeck et al. 2008). A fourth need, the “need for structure and safety” is added by some authors (e.g., Kasser 2009, 2011; Sheldon et al. 2001). This need is fulfilled in stable conditions of certainty about the future. It is threatened under conditions of uncertainty, e.g., when the future of one’s job is uncertain because of the company’s economic situation.

According to self-determination theory (e.g., Deci and Ryan 2000), opportunities to satisfy intrinsic needs facilitate self-motivation and effective functioning. In contrast, thwarted satisfaction of needs is assumed to undermine motivation and have maladaptive consequences. It is presumed that the fulfillment of basic needs is important for every employee. The strength of certain needs may differ at least to some degree across individuals (Van Yperen et al. 2014), but the fulfillment of basic needs is much more important than these minor differences between people in terms of individual needs. Thus, need satisfaction is more important than individual need strength. This has been confirmed by a number of studies with working populations. For instance, Van den Broeck et al. (2008) were able to show that psychological need satisfaction fully mediates the relations between job resources and outcome variables and at least partly mediates the relations between job demands and outcomes. Other studies confirmed the important role of need satisfaction for positive HR management outcomes (e.g., learning results; Marescaux et al. 2013) and work performance (Baard et al. 2004). Thus, the satisfaction of basic needs may serve as a useful evaluation criterion for assessing the role of new demands at work and implementing job design measures related to these new demands.

But what are the most important new demands? As shown in the previous chapters, there are three types of new demands which are confirmed to have additional effects as compared to conventional demands. *Work intensification* describes a new demand characterized by an intensification of work effort (Kubicek et al. 2015; Kubicek and Korunka in print; see also Paškvan and Kubicek, Chap. 3). Over many years, empirical studies (e.g., the European Working Conditions Surveys; Eurofound 2015) have confirmed that in many jobs, more and more tasks now have to be accomplished within one working day. People are required to work faster and conduct a number of different tasks at the same time. *Intensified planning and decision-making demands* refer to the fact that many jobs nowadays are characterized by an (ever) increasing amount of autonomy. Many jobs are no longer limited to specific work spaces or defined working times. In "boundaryless" jobs (e.g., Wohlers and Hertel in print), employees are able to work anywhere and at any time. Management by objectives (Drucker 1954) often defines clear targets but leaves how to reach these targets completely up to employees. Autonomy, which was previously seen only as a very important job resource (e.g., Karasek 1979), is gradually being converted into a new demand, at both the job and career levels. Finally, *intensified learning demands* describes the need to adapt to new working conditions triggered by an increased number of organizational change processes and the need to adapt to new information and communication technologies.

When thinking about job design, one has to ask what are the specific challenges related to the new demands. Different demands affect individuals by way of similar mechanisms (JD-R model; Bakker and Demerouti 2007). On the one hand, demands trigger the use of energy, which may lead to exhaustion, fatigue and reduced well-being. On the other hand, demands may also have positive motivational impacts, leading to increases in work engagement. Based on the challenge/hindrance approach (Cavanaugh et al. 2000), work intensification can be described as a hindrance demand, which requires an additional amount of energy consumption. On the other hand, intensified planning and decision-making demands and intensified learning demands are challenge demands, characterized both by increased energy use and positive motivational effects. However, new demands cannot be just generally classified as hindrance or challenge demands, assuming that they affect all employees in the same way. The current world of work also calls for increased individual self-regulation processes (see also Flecker et al., Chap. 2). Employees may differ in their self-regulation capabilities (Muraven and Baumeister 2000). Thus, individual appraisals of the challenging and/or hindering characteristics of demands (Widmer et al. 2012) have to be taken into consideration as well.

The following chapter describes the role of classic job design approaches in the context of new job demands and against the background of the aforementioned characteristics of these demands.

8.5 The Important Role of (Classic) Job Design in a Changing World of Work

Even if a changing world of work challenges classic job design in many ways (e.g., via the increasing complexity of job aspects, or differences in individual needs), classic job design measures are still relevant. The following chapter gives an overview of classic job design measures and their important role in a changing world of work.

Many job design theories focus mainly on the motivational aspects of work. One of the most influential motivational job design theories is Hackman and Oldham's job characteristics model (1976), which focuses on five general aspects of job design: autonomy, skill variety, task identity, task significance, and feedback. Humphrey et al. (2007) expanded the job characteristics model by adding further job and motivational characteristics. This expanded model was confirmed by meta-analytical data. Here, I use the expanded model of Humphrey et al. (2007) as a framework for evaluating the role of certain job design characteristics in a changing world of work. Humphrey et al. (2007) describe contextual characteristics, job stressors, and social and motivational characteristics in their expanded work.

8.5.1 The Unaltered View on Stressors in the Work Context

Physical demands refer to the amount of physical effort and physical activity related to the performance of tasks. In general, a decrease in physical demands is observable in the current world of work. Physical demands started to decrease a few decades ago with the transformation from production work to service work. Since then, many workplaces, especially in the field of production work, have been further improved, which led to a further reduction of physical demands. Automation processes and the use of new technologies have supported these developments, with many production workplaces now having relatively few physical demands. However, one should keep in mind that even an extremely low level of physical demands may have negative effects on employees' long-term health. Consider, for instance, monitoring workplaces where workers observe a number of control panels on screens with high concentration and little opportunity for physical exercise. General increases in terms of working hours and reduced breaks may increase such negative effects. Work intensification as a new demand may also lead to increased physical demands in certain workplaces where physical demands were already relatively high. This could be the case for instance in eldercare, where more people need to be served in shorter time intervals. An increase of health-related problems and performance reductions may be the result. Thus, the general reduction of physical demands notwithstanding, one still needs to test for potential increases in physical demands, especially in fields where a certain level of physical demands can be expected.

External working conditions are those aspects of the work environment which potentially affect employees' health like temperature, light, air quality, and noise. All of these conditions are important per se as they may strongly affect not only health but also productivity and quality of working life in general. However, even if intensified job demands exist widely independently of the working environment, they might still have something of an effect on employees' working conditions. For instance, work intensification may lead to increased exposure to unfavorable working conditions because of the need for prolonged working hours. On the other hand, increased planning and decision-making demands and the independence of work from specific work locations give employees the opportunity to select work places with favorable working conditions. Thus, it can be said that external working conditions need to be taken into consideration in exactly the same way as a few years ago.

Work interruptions are a well-researched important stressor leading not only to reduced productivity, but also to decreases in well-being and job satisfaction (Frese and Zapf 1994). Work interruptions occur at least to a certain degree in every workplace. Modern workplaces with strong ICT support have increased potential for work interruptions. New emails pop up constantly, disrupting a concentrated flow of work. In addition to the telephone, there are many other personal communication channels available which may also disturb our work. Increased self-control efforts are needed to effectively cope with work interruptions, since in many circumstances workers need to decide on their own if and when they want to check their emails or talk to other people. Such self-control efforts expend energy, which, in addition to increased energy consumption because of work intensification, may lead to reduced well-being (e.g., Schmidt and Diestel 2015). Thus, work interruptions should be also reduced by job and organizational design measures.

Excessive work load and time pressure are well-known classic stressors at work (e.g., Karasek 1979). These stressors have clear effects on the quality of working life. Excessive amounts of work load and time pressure need to be identified, and job design measures should be implemented to reduce these stressors. This becomes even more important in the current world of work, since work intensification – as a new job demand – has been found to have additional negative effects on the quality of working life (Kubicek et al. 2015). A challenge in this context is the fact that time pressure and work intensification should be treated separately. While a certain amount of time pressure and work load are inherent characteristics of any workplace (and, if not too high, not necessarily problematic from a job design perspective per se), work intensification should definitely be considered a relevant stressor (Franke 2015).

Another important job design aspect involves *ergonomic aspects of work*. The ergonomic design of work serves as a foundation not only for high productivity ("Good ergonomics is good economics"; Hendricks 2003), but also for a high quality of working life. Good ergonomic design should be a matter of course for any workplace. Conventional workplaces are usually located at one fixed place. But with the increasing prevalence of boundaryless work settings, an increasing number of workplaces exist, as employees have the opportunity to work from different locations both within and outside the organization and from home. Measures to improve ergonomic aspects of work are usually limited to workplaces within the organization. Employees often do not think about ergonomics when working from home. This, of course, is another challenge for work design in the current world of work.

A good workplace is also a safe workplace. *Workplace safety* has significantly improved over the last decades. Decreasing accident rates can be observed for all countries in the Western world (e.g., Eurostat 2015). But we should not take these positive developments for granted. Increases in working speed, time pressure and of course work intensification are definitely risk factors for workplace safety (e.g., Brenner et al. 2004). Lack of attention is a well-confirmed cause of accidents, and this risk factor increases with high amounts of these stressors. Although these risks are not yet visible in accident rates, these rates can be expected to increase again over the next years. It may even be the case that these factors already play an important role in accidents, but their negative effects are hidden by concurrent continuous improvements in safety technologies. Thus, a further decrease in accidents should not be taken for granted. It will be one of the big challenges for work design in the next years.

8.5.1.1 Motivational Aspects of Work Design: Optimization Instead of Maximization

Hackman and Oldham (1976) described five motivational task characteristics which have been conclusively confirmed by many empirical studies. These five characteristics, namely autonomy, skill variety, task identity, task significance and feedback, are thought to impact critical psychological states. Skill variety, task identity and task significance are related to the experience of meaningfulness, autonomy is related to perceived responsibility, and feedback is related to knowledge of results. These critical psychological states, in turn, lead to positive behavioral (performance), attitudinal (job satisfaction), and well-being (low stress) outcomes. All of these motivational characteristics are still relevant job design elements in the context of new job demands and new forms of work.

Skill variety refers to the extent to which a job requires a person to utilize different skills. In general, one may assume that skill variety is relevant in all types of jobs, and also in new forms of work. It is obvious that the extreme of very low skill variety (for instance, very monotonous and repetitive work) has negative effects on psychological states and on outcomes. On the other hand, intensified job demands may be related to increases in skill variety. Work intensification may lead to the need to perform additional tasks. Intensified learning demands are an expression of the need for employees to learn new things and perform new tasks. Thus, job design focusing on increasing skill demands (job enlargement) is still relevant for many jobs. On the other hand, it needs to be taken into consideration that a significant intensification of job demands may be related to increases in skill variety. For instance, strong increases in autonomy in the form of job expansion measures may also strongly increase the number of different tasks that have to be performed. Thus, at least in certain circumstances, job design should target at the *right amount* of skill variety instead of the maximum amount.

Task identity refers to the degree to which a single person is in charge of completing an identifiable piece of work from start to finish. Low levels of task identity are observable when tasks are fragmentary or incomplete. Tayloristic work design is characterized by very low levels of task identity. Whereas Tayloristic work design may be seen as a historical artifact, a comeback of Tayloristic principles can be observed in some areas (e.g., Kempe 2001). Work in call centers is one example of a relatively new area of work usually organized on Tayloristic principles with low levels of task identity. On the other hand, new forms of work and intensified job demands are not necessarily related to changes in task identity. Intensified job demands may be observable in jobs where task identity remains unchanged. It may be even the case that intensified job demands can go hand in hand with reductions in task identity. Thus, designing jobs with a high level of task identity or improving jobs by increasing task identity are still important goals for work design in the current world of work.

Task significance refers to whether a certain job substantially affects other people's work, health or well-being. High task significance is especially important for perceptions of meaning in jobs. Task significance is a core job characteristic, relatively independent from the form of work and unrelated to intensified job demands. If there were to be a relation between intensified job demands and task significance, one might expect this relation to be positive. For instance, it may be the case that intensified learning demands are related to increases in task significance. Thus, task significance as a core element of job design needs to be taken into consideration when designing jobs, relatively independently of intensified job demands.

Feedback refers to the degree to which people are able to recognize how effective they are at work. Feedback at work may stem from other people, such as supervisors or coworkers, but can also be contained within certain job tasks, for instance in direct customer interactions. Feedback is an essential element of job design. A job without feedback provides no motivation. And again, changes in feedback are to be expected when new forms of work are implemented and/or intensified job demands are observable. Feedback from the job itself does not necessarily change when jobs become intensified. On the other hand, work intensification may leave less time for personal feedback from supervisors or coworkers. Many studies have confirmed (e.g., Humphrey et al. 2007) that personal feedback may be more important than feedback from the job itself. Thus, in new forms of work under intensified working conditions, it is important to guarantee a high level of feedback. Guaranteeing a certain level of personal feedback might be particularly challenging in boundaryless jobs because supervisors or coworkers may not be available when needed.

In the original model of work motivation (Hackman and Oldham 1976), *work autonomy* was described as a one-dimensional construct with general positive effects on productivity, job satisfaction, and well-being. However, soon after this original description, many scholars suggested that instead of a one-dimensional construct, autonomy has different facets. Jackson et al. (1993) distinguished between work scheduling autonomy and methods autonomy. The former refers to temporal aspects of autonomy, like the freedom to control the scheduling and timing of work, whereas the latter refers to the freedom of control which methods and tools are used in a certain workplace. Another aspect of autonomy is decision-making autonomy, which refers to the freedom to make work-related decisions. In general, autonomy has been found to be related to both performance and job satisfaction and to improved well-being. However, certain effects of each of the three facets of autonomy can be expected. For instance, employees with a high degree of work methods autonomy should be able to exert greater influence on how a certain task is to be accomplished, whereas work scheduling autonomy in relation to a certain task is simply related to the fact that the order of specific behaviors may be changed.

New demands at work, especially intensified planning and decision-making demands, are definitely related to increases in autonomy. At least two of the three facets of autonomy should increase. Intensified planning and decision-making demands are related to increases in both decision-making autonomy and time scheduling autonomy. As stated in more detail in Chap. 4, there are not only "bright" sides but also "dark" sides to autonomy. Especially high levels of autonomy may lead to negative effects for employees, similar to the effects of reduced autonomy (Kubicek et al. 2014). In particular, highly flexible settings with regard to worktime and workplace may bring about negative effects (Baltes et al. 1999). Based on Ryan and Deci's self-determination theory, there exists a general need for autonomy, but the strength of this need may differ somewhat between people (Van Yperen et al. 2014). Employees with a high need for autonomy may also benefit from work settings with very high amounts of autonomy, like fully boundaryless work with the opportunity to perform work tasks anywhere and at any time. Thus, job design should take individual differences in need for autonomy into consideration at least to some degree. In some cases, it might even be better to reduce extreme amounts of autonomy via job design interventions. For instance, in boundaryless work settings, at least some "boundaries", e.g., an agreement on times in which employees are not expected to work, could be introduced.

8.6 Social Characteristics: Virtualization Versus Personal Contacts

Individual jobs are embedded in organizational and social contexts. These contexts not only affect work design interventions at the job level (e.g., Humphrey et al. 2007), but are also important focal points for workplace interventions.

The social context of work is defined as the interpersonal interactions and relationships that are embedded in and influenced by the jobs, roles, and tasks that employees perform and enact (Grant et al. 2010). One of the most important social characteristics of work is *social support*, defined as the degree to which employees receive assistance from supervisors and coworkers (Karasek 1979). Social support may be perceived as emotional support, cognitive support, and instrumental support (House 1981).

In intensified working environments, all three forms of social support remain important or even gain increased relevance compared to conventional workplaces. Social support helps in coping with phases of work intensification. In rapidly changing working environments, instrumental social support is essential (see

Paškvan and Kubicek, Chap. 3). In boundaryless work settings, direct social interaction is often dramatically reduced or even nonexistent. In such settings, a work design intervention could involve arranging social meetings at certain points in time, e.g., a weekly team meeting with a commitment to personal attendance.

Morgeson and Humphrey (2006) described task interdependence as another important social characteristic. Task interdependence is the degree to which an employee's job is connected to other jobs, such that employees rely on each other to complete tasks. Highly interdependent jobs provide a base for intense personal contact and more opportunities to communicate with other employees. Such contacts help employees develop individual roles and to clarify personal role concepts. Task interdependence also serves as a basis for personal feedback. On the other hand, work intensification may further increase task interdependence. Again, in job design interventions one should keep in mind that very high levels of certain job features may have even negative effects (Warr 2009). Thus, job design interventions should be based on the recognition of the right amount of resources.

Another social characteristic mentioned by Morgeson and Humphrey (2006) is interactions outside the organization, defined as the extent to which a job requires an employee to communicate with people (customers, suppliers etc.) outside of the organization. Again, such contacts may serve as a base for communication, social support and feedback. Here again, personal contacts are more important than contacts via electronic communication.

A somewhat different social dimension of job design is *perceived organizational support*, defined as employees' perceptions regarding the extent to which the organization values their contributions and cares about their well-being. Employees who experience high amounts of perceived organizational support feel safe in the knowledge that help will be available from the organization when needed to carry out their job effectively and especially that support will be there in situations involving high stress and workload. Perceptions of organizational support are related to increases in well-being, better job satisfaction, and reduced strain. According to social exchange theory (for an overview see Cropanzano and Mitchell 2005), high productivity can also be expected when perceived organizational support is high. Rhoades and Eisenberger (2002) describe three important antecedents of perceived organizational support: fairness, supervisor support and organizational rewards.

Supervisor support as a form of social support is defined as the degree to which a supervisor values an employee's contributions and cares about his/her well-being. Since supervisors are often viewed as agents of the organization, supervisor support is perceived as an expression of organizational support.

Fairness relates to procedural and distributional aspects (Greenberg and McCarty 1990). Procedural aspects of fairness may be further distinguished into structural and social aspects (Cropanzano and Greenberg 1997). Structural aspects refer, for example, to the existence of formal rules and policies concerning decisions related to the employee. Some of the most important dimensions in this regard are the perception that information is received accurately and timely, and voice (employee input in decision-making processes). Social aspects of fairness refer to the perception that one is being treated with dignity and respect and the provision of information about the evaluation of outcomes and personal productivity.

Another antecedent of perceived organizational support is *organizational rewards*, which is per se a multi-faced construct. Fairness of pay and promotions are core aspects of organizational rewards. Another important aspect is *job security*. As a result of the economic turmoil of recent years, this aspect has become even more important over time. Many organizations are no longer able to guarantee their employees a certain basic amount of job security. Thus, high job security is experienced as an especially strong sign of organizational support.

Participative climate is another well-researched organizational design characteristic. It has many positive effects on productivity and quality of working life. It supports employees' empowerment (Spreitzer 1996). This may be even more important in the current world of work, where a return of Tayloristic principles is observable and new challenges for individuals occur.

8.7 Subjectification: New Challenges for Individuals

"Subjectification" of work refers to the fact that there is an ever-increasing tendency for workers to bring their personal views, aspirations and subjective standards into their work. In practical terms, subjectification means that companies have begun loading the uncertainties of the market onto the individual worker (Hurtienne et al. 2014), leading to higher levels of job, worktime and workplace autonomy (Wood and de Menezes 2011). "Boundaryless" work is characterized by the fact that work can be done anywhere and at any time. This new type of work does not only result in new challenges for job design, but also for individuals. Increased self-control and self-management are needed to deal with the heavy increase in planning and decision-making demands. Employees who consistently and carefully plan their goals and actions and who are able to set clear work boundaries by actively and continuously deciding where and when to work not only to reach their work goals but also save personal resources and are better able to deal with new demands. On the other hand, if individual self-control is low and employees are not able to utilize efficient time management strategies, the demands of subjectification and boundaryless work may lead to increases in strain and reduced work productivity. Thus, the acquisition of self-control and time management strategies is a successful individual strategy for preventing such negative effects of work. Organizations could support individuals by offering time management training for their employees. These trainings should also support individuals in developing personal boundary management strategies. These could include, for example, a clear commitment to non-work times, for instance on the weekend, a personal work hour limit, or a decision to confine work to certain places. The application of these strategies helps build up psychological detachment, which, in turn, buffers the negative effects of job stressors on individuals (e.g., Sonnentag and Fritz 2015). The positive effects of successful psychological detachment have been widely confirmed

(Sonnentag and Fritz 2015). Personal coping strategies help to improve psychological detachment from work. For instance, a high capacity for self-regulation (Muraven et al. 1999), clear segmentation preferences (Kreiner 2006), and high levels of self-efficacy support the ability to detach. From a more general perspective, individuals with positive psychological capital ("PSYCAP", Luthans and Youssef 2004), consisting of the four components hope, self-efficacy, optimism, and resilience, should be able to better cope with the demands of the current world of work. While these individual factors clearly help people cope with work demands, the main responsibility for job design that supports employees in dealing with new demands remains within the organization and even within society.

8.8 Macro-Level Considerations: Information and Political Strategies Are Becoming More Important

As described in more detail in the previous chapters, new job demands are a result of complex societal trends that are leading to fundamental changes in the organization of work. The shift from production to service work was probably the first and most important trigger of such changes. Globalization and increased competition are affecting the work force in many countries. The rapid development of new information and communication technologies is a strong driver of these changes. All of these developments have led to a series of changes in contemporary working life such as around-the-clock availability, massive changes in working hours, temporary jobs, uncertain employment relations, and – above all – to new job demands as described in these chapters.

In the Western world, there is a general acknowledgement of such developments in the world of work. For instance, the European Union has responded to changes in the world of work formulating EU directives. These directives are legally binding, and member countries are asked to transpose the directives into national laws. Especially important for the current topic was the EU framework directive of June 12, 1989 on the "Introduction of measures to encourage improvements in the safety and health of workers in the European Union". This framework directive has a wide scope and formulates both minimum requirements and fundamental principles, for instance the principle of prevention and risk assessment, and defines the responsibilities of employers and employees. The directive obliges all employers in Europe to assess occupational safety and health risks that could potentially negatively influence the health of their employees. Since all potential risks have to be assessed, psychosocial risks, and also new demands which may affect the health of the employees, are included. Furthermore, the framework directive lays out a stepwise approach for dealing with risk factors. First, the potential risks have to be evaluated. Employers need to find out whether a risk can be eliminated or, if not fully eliminated, at least be reduced via organizational measures. Only after implementing or strengthening such measures should further measures be taken to strengthen the individual. As an example of an implementation of the EU framework directive, the following section describes the evaluation of psychosocial risks in Austria (see Korunka 2013):

The evaluation of psychosocial risks at workplaces in Austria is based on the regulations of the European Union. Since January 2103 and based on similar laws in the other German speaking countries, employers have been requested to evaluate psychosocial risks of their employees (Federal Ministry of Labor, Social Affairs and Consumer Protection 2012). The evaluation of workplaces includes the identification of psychosocial risks, the evaluation of potential dangers, the development of measures to reduce and prevent risk, and the documentation of the evaluation process. As in other European countries, the evaluation is based on the International Standard ISO EN 10075 (Ergonomic principles related to mental workload). To fulfill the criteria defined by the law, the Ministry of Labor recommends the following evaluation process for companies:

- *Planning and development of an action plan*: First, a steering committee should be established, including internal and external experts (work and organizational psychologists, health and safety experts, specialists in occupational health medicine). Next, existing information (e.g., existing evaluations of workplaces, health related projects etc.) should be collected. Critical areas of work and specific psychosocial risks need to be defined. Based on these facts, a process for evaluating psychosocial risks should be planned and information should be provided to employees.
- *Evaluation of psychosocial risks*: Adequate and standardized instruments should be used for the evaluation. Depending on the specific context, the use of standardized questionnaires, group interviews, single interviews, focus groups and expert observations are recommended.
- *Development of measures*: Based on the evaluation results, a process for developing intervention measures should be established. Health and safety experts, members of the Labor Union, and employees should participate in this process.
- *Implementation of measures and quality control*: The measures should be implemented and their success should be evaluated after a certain time period. If signs of increases in psychosocial risks (e.g., complaints, health issues, accidents etc.) are observable, a further evaluation is necessary.
- *Documentation*: The evaluation of psychosocial risks and the development of intervention measures need to be documented. Employees should be informed about their psychosocial risks and measures to improve them.

An important characteristic of the evaluation of psychosocial risks is the fact that risk evaluation and the development of measures should focus on the causes and sources of the problem, not on employees' behavior or satisfaction. The following four dimensions of sources of psychosocial risks should be included in the evaluations:

- *Task- and job-immanent psychosocial risks*: Work load, physical load, cognitive load, emotional demands, qualification problems, etc.
- *Psychosocial risks on the organizational level*: Problems in collaboration and teamwork, information deficits, autonomy deficits, etc.
- *Psychosocial risks of the work environment*: Climate, noise, light, working space, work safety, etc.
- *Psychosocial risks on the level of work organization*: Multitasking, goal conflicts, a lack of goals and priorities, work interruptions, prolonged work time, a lack of relaxation times, time pressure, work load, etc.

Since the implementation of the new law in January 2013, the Austrian Labor Inspectorate has urged companies to evaluate their employees' psychosocial risks. Many evaluation processes have been started, and experience in implementing the new law has been gained.

Based on the European framework directive, other countries have developed different approaches with the goal of reducing psychosocial risks. For instance, Denmark implemented a "national strategy for improving the working environment" in 2010 (Peiro and Molina 2013). New demands and the psychosocial working environment play a prominent role in this strategy as burnout and stress are acknowledged as serious health and safety issues. The Danish strategy does not just emphasize that a good working environment has to be ensured at the organizational level, but also that the psychosocial working environment has to be considered when conducting organizational changes.

8.9 Final Thoughts

From a job design perspective, a good job does not just combine a high quality of working life and productivity. It also maintains the jobholder's health and supports his/her personality development (Hacker 2005). Work and organizational psychology has accumulated a multitude of research clearly describing the characteristics of such a job. These characteristics of job design, as also presented in this chapter, still apply in the current world of work. Basic elements and core ideas of job design, like the reduction of physical demands, increasing workplace safety, the creation of supportive working conditions that help to maintain health, and the improvement of ergonomic work designs will always be important, just as the well-confirmed motivational qualities of good jobs are unlikely to change over time. On the other hand, certain developments in the current world of work, like increasing opportunities to work anywhere and at any time, alter the goals of job design measures to a certain degree. Instead of maximizing any given feature of a job, optimizing this feature should be the goal, at least for certain characteristics, like autonomy or learning demands. Individuals' needs should be considered when formulating optimization targets.

The question remains whether as the considerations presented in this chapter will also apply to work in the future. Looking at the rapid speed of change observable in the last few decades, many further changes and developments can be expected in the near future.

However, it is not easy to predict what changes exactly the future will bring for the world of work. The rapid development of ICT will certainly continue over the coming decades, and this will lead to changes and developments that are not fully predictable at the present time. A good example of this is the Internet of Things (Industry 4.0), currently the subject of widespread public discussion but barely realized in the world of work. With a full implementation of Industry 4.0, some experts predict many improvements for employees. Other researchers point to the potential for huge changes in qualification demands and a (further) polarization of work, including a strong increase in unemployment. Such developments which are hard to predict will pose a challenge for research in work and organizational psychology in the near future.

References

Baard, P. P., Deci, E. L., & Ryan, R. M. (2004). Intrinsic need satisfaction: A motivational basis of performance and weil-being in two work settings. *Journal of Applied Social Psychology, 34,* 2045–2068. doi:10.1111/j.1559-1816.2004.tb02690.x.

Bakker, A. B., & Demerouti, E. (2007). The job demands-resources model: State of the art. *Journal of Managerial psychology, 22,* 309–328. doi:10.1108/02683940710733115.

Baltes, B. B., Brigs, T. E., Huff, J. W., Wright, J. A., & Neuman, G. A. (1999). Flexible and compressed workweek schedules: A meta-analysis of their effects on work-related criteria. *Journal of Applied Psychology, 4,* 496–513.

Barley, S. R., Grobal, S., & Meyerson, D. E. (2011). E-mail as a source and symbol of stress. *Journal of Organization Science, 22*(4), 887–906.

Brenner, M. D., Fairris, D., & Ruser, J. (2004). "Flexible" work practices and occupational safety and health: Exploring the relationship between cumulative trauma disorders and workplace transformations. *Industrial Relations, 43,* 242–266.

Brod, C. (1984). *Technostress: The human cost of the computer revolution*. Addison Wesley Publishing.

Carayon, P., & Smith, M. J. (2014). The balance concept revisited: Finding balance to reduce stress in a frantic world of IT. In C. Korunka & P. Hoonakker (Eds.), *The impact of ICT in quality of working life* (pp. 105–121). Amsterdam/New York: Springer.

Cavanaugh, M. A., Boswell, W. R., Roehling, M. V., & Boudreau, J. W. (2000). An empirical examination of self-reported work stress among U.S. managers. *Journal of Applied Psychology, 85,* 65–74. doi:10.1037/0021-9010.85.1.65.

Chesley, N. (2014). Information and communication technology use, work intensification and employee strain and distress. *Work, Employment & Society, 28,* 589–610. doi:10.1177/0950017013500112.

Cropanzano, R., & Greenberg, J. (1997). Progress in organizational justice: Tunneling through the maze. *International Review of Industrial and Organizational Psychology, 12,* 317–372.

Cropanzano, R., & Mitchell, M. S. (2005). Social exchange theory: An interdisciplinary review. *Journal of Management, 31,* 874–900. doi:10.1177/0149206305279602.

Deci, E. L., & Ryan, R. M. (2000). The "what" and "why" of goal pursuits: Human needs and the self-determination of behavior. *Psychological Inquiry, 11,* 227–268. doi:10.1207/S15327965PLI1104_01.

Drucker, P. F. (1954). *The practice of management*. New York: Harper & Row.
Eurofound. (2015). *Sixth European working conditions survey*. Retrieved from http://www.eurofound.europa.eu/de/surveys/2015/sixth-european-working-conditions-survey-2015
Eurostat. (2015). *Accident at work statistics*. Retrieved from http://ec.europa.eu/eurostat/statistics-explained/index.php/Accidents_at_work_statistics/de
Franke, F. (2015). Is work intensification extra stress? *Journal of Personnel Psychology, 14,* 17–27.
Frese, M., & Zapf, D. (1994). Action as the core of work psychology: A German approach. In H. C. Triandis & M. D. Dunnette (Eds.), *Handbook of industrial and organizational psychology* (2nd ed., pp. 271–340). Palo Alto, California.
Future Work Centre. (2016). *You've got mail*. Retrieved from http://www.futureworkcentre.com/what-we-do/insight/email-at-work/
Grant, A. M., Fried, Y., Parker, S. K., & Frese, M. (2010). Putting job design in context: Introduction to the special issue. *Journal of Organizational Behavior, 31,* 145–157. doi:10.1002/job.679.
Greenberg, J., & McCarty, C. (1990). The interpersonal aspects of procedural justice: A new perspective on pay fairness. *Labor Law Journal, 41,* 580.
Hacker, W. (2005). *Allgemeine Arbeitspsychologie [General Work Psychology]*. Bern: Huber publishers.
Hackman, J. R., & Oldham, G. R. (1976). Motivation through the design of work: Test of a theory. *Organizational Behavior and Human Performance, 16,* 250–279.
Hendrick, H. W. (2003). Determining the cost–benefits of ergonomics projects and factors that lead to their success. *Applied Ergonomics, 34,* 419–427. doi:10.1016/S0003-6870(03)00062-0.
House, J. S. (1981). *Work stress and social support*. Addison-Wesley Pub. Co.
Humphrey, S. E., Nahrgang, J. D., & Morgeson, F. P. (2007). Integrating motivational, social, and contextual work design features: A meta-analytic summary and theoretical extension of the work design literature. *Journal of Applied Psychology, 92,* 1332–1356. doi:10.1037/0021-9010.92.5.1332.
Hurtienne, J., Stilijanow, U., & Junghanns, G. (2014). Time and work pressure in today's working world. In C. Korunka & P. Hoonakker (Eds.). *The impact of ICT in quality of working life* (pp. 63–86). New York, Amsterdam: Springer.
Jackson, P. R., Wall, T. D., Martin, R., & Davids, K. (1993). New measures of job control, cognitive demand, and production responsibility. *Journal of Applied Psychology, 78,* 753–762.
Karasek, R. A., Jr. (1979). Job demands, job decision latitude, and mental strain: Implications for job redesign. *Administrative Science Quarterly*, 285–308. doi:10.2307/2392498
Kasser, T. (2009). Psychological need satisfaction, personal well-being, and ecological sustainability. *Ecopsychology, 1,* 175–180. doi:10.1089/eco.2009.0025.
Kasser, T. (2011). Can thrift bring well-being? A review of the research and a tentative theory. *Social and Personality Psychology Compass, 5,* 865–877. doi:10.1111/j.1751-9004.2011.00396.x.
Kempe, M. (2001). A comeback for "Taylorism"? Reposts from Germany. *World of Work, 38,* 18–20.
Korunka, C. (2013). The evaluation of psychosocial risks. Case studies and experiences from Austria. In J. M. Peiro & C. Molina (Eds.), *International yearbook on psychosocial risk prevention and quality of life at work 2013.*
Korunka, C., & Hoonakker, P. (2014). The Future of ICT and Quality of Working Life: Challenges, Benefits, and Risks. In C. Korunka & P. Hoonakker (Eds.). *The Impact of ICT on Quality of Working Life* (pp. 205-219). New York, Amsterdam: Springer.
Kreiner, G. E. (2006). Consequences of work-home segmentation or integration: A person-environment fit perspective. *Journal of Organizational Behavior, 27,* 485–507. doi:10.1002/job.386.
Kubicek, B., Korunka, C., & Tement, S. (2014). Too much job control? Two studies on curvilinear relations between job control and eldercare workers' well-being. *International Journal of Nursing Studies, 51,* 1644–1653. doi:10.1016/j.ijnurstu.2014.05.005.

Kubicek, B., Paškvan, M., & Korunka, C. (2015). Development and validation of an instrument for assessing job demands arising from accelerated change: The intensification of job demands scale (IDS). *European Journal of Work and Organizational Psychology, 24,* 898–913.

Kubicek, B., & Korunka, C. (in print). Job demands in a changing world of work. In R. Burke & K. Page (Eds.), *Research handbook on work and wellbeing*. Cheltenham, UK: Edward Elgar Publishing.

Leather, P., Pyrgas, M., Beale, D., & Lawrence, C. (1998). Windows in the workplace sunlight, view, and occupational stress. *Environment and Behavior, 30,* 739–762. doi:10.1177/001391659803000601.

Luthans, F., & Youssef, C. M. (2004). Human, social, and now positive psychological capital management. *Organizational Dynamics, 33,* 143–160.

Marescaux, E., de Winne, S., & Sels, L. (2013). HR practices and HRM outcomes: The role of basic need satisfaction. *Personnel Review, 42,* 4–27. doi:10.1108/00483481311285200.

Morgeson, F. P., & Humphrey, S. E. (2006). The work design questionnaire (WDQ): Developing and validating a comprehensive measure for assessing job design and the nature of work. *Journal of Applied Psychology, 91,* 1321. doi:10.1037/0021-9010.91.6.1321.

Muraven, M., & Baumeister, R. F. (2000). Self-regulation and depletion of limited resources: Does self-control resemble a muscle? *Psychological Bulletin, 126,* 247–259. I0.I037//0033-2909.126.2.247.

Muraven, M., Baumeister, R. F., & Tice, D. M. (1999). Longitudinal improvement of self-regulation through practice: Building self-control strength through repeated exercise. *Journal of Social Psychology, 139,* 446–458.

Paškvan, M., Kubicek, B., Prem, R., & Korunka, C. (2015). Cognitive appraisal of work intensification. *International Journal of Stress Management*. doi:10.1037/a0039689.

Peiro, J. M., & Molina, C. (Eds.). (2013). *International yearbook on psychosocial risk prevention and quality of life at work 2013.*

Rhoades, L., & Eisenberger, R. (2002). Perceived organizational support: A review of the literature. *Journal of Applied Psychology, 87,* 698. doi:10.1037/0021-9010.87.4.698.

Rosa, H. (2005). Beschleunigung. *Die Veränderung der Zeitstrukturen in der Moderne*. Frankfurt am Main: Suhrkamp.

Ryan, R. M., & Deci, E. L. (2000). Self-determination theory and the facilitation of intrinsic motivation, social development, and well-being. *American Psychologist, 55,* 68–78. doi:10.1037/0003.066X.55.1.68.

Salanova, M., Llorens, S., & Cifre, E. (2013). The dark side of technologies: Technostress among users of information and communication technologies. *International Journal of Psychology, 48,* 422–436. doi:10.1080/00207594.2012.680460.

Schaufeli, W., Taris, T. W., & van Rhenen, W. (2008). Workaholism, burnout, and work engagement: Three of a kind or three different kinds of employee well-being? *Applied Psychology: An International Review, 57,* 173–203. doi:10.1111/j.1464-0597.2007.00285.x.

Schmidt, K. H., & Diestel, S. (2015). Self-control demands. *Journal of Personnel Psychology, 14,* 49–60. doi:10.1027/1866-5888/a000123.

Sheldon, K. M., Elliot, A. J., Kim, Y., & Kasser, T. (2001). What is satisfying about satisfying events? Testing 10 candidate psychological needs. *Journal of Personality and Social Psychology, 89,* 325–339. doi:10.1037/0022-3514.80.2.325.

Smith, M. J., & Sainfort, P. C. (1989). A balance theory of job design for stress reduction. *International Journal of Industrial Ergonomics, 4,* 67–79. doi:10.1016/0169-8141(89)90051-6.

Sonnentag, S., & Fritz, C. (2015). Recovery from job stress: The stressor-detachment model as an integrative framework. *Journal of Organizational Behavior, 36,* 72–103. doi:10.1002/job.1924.

Spreitzer, G. M. (1996). Social structural characteristics of psychological empowerment. *Academy of Management Journal, 39,* 483–504.

Ulich, E. (1978). Über das Prinzip der differentiellen Arbeitsgestaltung. *Industrielle Organisation, 47,* 566–568.

Van den Broeck, A., Vansteenkiste, M., & De Witte, H. (2008). Self-determination theory: A theoretical and empirical overview in occupational health psychology. In J. Houdmont &

S. Leka (Eds.), *Occupational health psychology: European perspectives on research, education, and practice* (pp. 63–88). Nottingham: Nottingham University Press.
Van Yperen, N. W., Rietzschel, E. F., & De Jonge, K. M. (2014). Blended working: For whom it may (not) work. *PLoS ONE, 9*(7), e102921. 0.1371/journal.pone.0102921.
Warr, P. (2009). Environmental "vitamins", personal judgments, work values, and happiness. In S. Cartwright & C. Cooper (Eds.), *The Oxford handbook of organizational well-being* (pp. 57–85). Oxford: Oxford University Press.
Widmer, P. S., Semmer, N. K., Kälin, W., Jacobshagen, N., & Meier, L. L. (2012). The ambivalence of challenge stressors: Time pressure associated with both negative and positive well-being. *Journal of Vocational Behavior, 80*, 422–422. doi::10.1016/j.jvb.2011.09.006
Wohlers, C., & Hertel, G. (in print). *Choosing where to work @work—towards a theoretical model explaining benefits and risks of flexible office concepts.*
Wood, S., & de Menezes, L. M. (2011). High involvement management, high performance work systems and well-being. *The International Journal of Human Resource Management, 22*, 1586–1610. doi:10.1080/09585192.2011.561967.

Author Biography

Christian Korunka is Professor of Work and Organizational Psychology at the Faculty of Psychology, University of Vienna. He has published journal articles and book chapters dealing with new job demands, organizational change processes, and quality of working life. His current research focuses on new forms of work (e.g., new office concepts, boundaryless work), new job demands and their role in shaping quality of working life. Another stream of his research deals with psychological aspects of entrepreneurship and family businesses.

Chapter 9
The Present and Future of Work: Some Concluding Remarks and Reflections on Upcoming Trends

Bettina Kubicek and Christian Korunka

The chapters of this book have outlined developments in the organization of work as well as in work and employment conditions and analyzed their likely effects on employees' well-being, motivation and career prospects. The authors have tried to provide a picture of how socio-economic changes increase work intensity (Paškvan and Kubicek, Chap. 3) and autonomy (Kubicek et al., Chap. 4), shift career responsibility from organizations to employees (Nalis, Chap. 5) and make work more flexible (Gerdenitsch, Chap. 6) as well as of the positive and negative consequences these changes may lead to. Moreover, they discussed potential underlying mechanisms explaining these effects (see especially Prem, Chap. 7) and consequences for job redesign approaches (Korunka, Chap. 8). The authors have taken critical as well as affirmative outlooks on the reorganization of work. While Paškvan and Kubicek, (Chap. 3), Kubicek et al. (Chap. 4) and Nalis (Chap. 5) highlighted potential demands stemming from increasing work intensity as well as increasing job-related and career-related planning and decision-making demands, Gerdenitsch (Chap. 6) underscored the potential of flexible work arrangements (i.e. temporally, spatially flexible work and nomadic work) to satisfy employees' basic human needs.

Despite these opposing evaluations of potential consequences of the current changes in working conditions for employees, we argue that there are certain comprehensive patterns characterizing the reorganization of work. One pattern involves the intensification of work, which puts more pressure on employees (i.e., increasing work intensity), while a second pattern involves the deregulation of work, which increases employees' responsibility for their work and careers as well as employees' temporal and spatial flexibility. A third pattern involves the subjectification of work (i.e., transformations of employees' autonomy at work). In this concluding chapter, we will reflect on these patterns as well as on potential dividing

B. Kubicek (✉) · C. Korunka
Faculty of Psychology, Universitätsstraße 7, 1010 Vienna, Austria
e-mail: bettina.kubicek@univie.ac.at

C. Korunka and B. Kubicek (eds.), *Job Demands in a Changing World of Work*,
DOI 10.1007/978-3-319-54678-0_9

lines and bring in some reflections on future developments and upcoming challenges. Our overall assumption is that current trends will create new dividing lines among employees, with individual abilities and skills now playing a greater role than previously, and that the effects of coming trends will unfold along similar lines as current developments and fit into the patterns of the intensification, deregulation and subjectification of work.

9.1 The Reorganization of Work: Intensification, Deregulation and Subjectification

One comprehensive pattern relates to quantitative changes in the demands put on employees. Work in many areas has intensified, and continuous restructuring, which is supposed to improve efficiency, has added to employees' workloads. The emergence of flexible work scheduling (Brewer 2000), new information and communication technologies, and the decentralization of top management control (Cascio 2003) have brought about increases in autonomy for many employees (Wood 2011), accompanied by a greater emphasis on speed and timeliness (Garhammer 2002). In order to deal effectively with higher goals and tighter deadlines, employees are given substantial discretion in deciding when and where to work as well as in deciding which tasks to pursue and which methods, materials, and tools to use (Tripoli 1998). Although these developments bring about more freedom and decision latitude for employees, they might also further increase employees' workload due to the fact that previously predetermined work settings need to be arranged by employees themselves. In addition to meeting higher goal expectations regarding their core tasks, employees are required to organize their work and their collaboration with coworkers, supervise goal achievement, and take on duties previously performed by line managers without being equipped with appropriate resources (including, for example, administrative staff). Greater autonomy might also encourage employees to put more effort into their job, which contributes to work intensification (see e.g., Kelliher and Anderson 2010).

These developments already hint at a second pattern of changes, namely the deregulation of work and its accompanying demands for individual workers. With the aim of making work more flexible, rules and regulations have been softened or completely abolished, increasing workers' individual responsibility with regard to how perform their jobs. Additionally, we have witnessed an increase in flexible work arrangements. Work no longer needs to be performed within specific working hours or at a specific workplace but can be carried out when and where it best suits the task or the person performing it. This not only entails increased freedom for employees but also brings about obligations to actually plan, structure and coordinate one's work with others (Pongratz and Voß 2003). In fact, the regulation of work has been handed over to the individual, who takes personal responsibility for his/her work (Allvin et al. 2011). But it is not only the rules and regulations regarding work itself that have been weakened or even entirely abolished; this has also been the case for the regulation of organizational careers and employment. As pointed out by Irina Nalis in Chap. 5, responsibility for developing employees' careers has shifted from the organization to the individual. Thus, many employees are increasingly required to manage their careers themselves and to maintain their employability inside and outside their current organization. Although affirmative outlooks on this development prevail in the literature, Nalis also discussed its potential downsides (see Baruch and Vardi 2015 for a critical review), such as organizations' decreasing willingness to invest in employees' knowledge and skills and the accompanying negative effects on employees' employability.

Although organizations are more reluctant to invest in employees' careers, they are simultaneously seeking to more extensively appropriate employees' personal resources (e.g., self-organization, creativity, sense of responsibility, etc.). At the same time, employees are increasingly willing to bring their subjectivity to work (e.g., opinions and motivations), to apply and further develop their skills and to actively shape their work situation (see, for instance, the growing literature on job crafting). This process of increasing interdependence between individuals and their work is termed subjectification (Pongratz and Voß 2003) and may be seen as a third pattern of changes linked to the intensification and deregulation of work. In order to make use of employees' subjective potential and to react to individuals' need to bring their subjectivity to work, organizations have increased employees' discretion in terms of how, when and where work is conducted (Paškvan and Kubicek, Chap. 3). Yet this trend is far from unambiguously positive. The ambivalence of work subjectification is particularly obvious in today's world of work, in which creativity as well as autonomous and responsible action and decision-making are not only allowed but required of employees (Weiskopf and Loacker 2006).

As this short account highlights, the three patterns of change identified here are intertwined and are likely to reinforce each other. As the need to provide goods and services just in time increases, organizations are trying to make work more efficient. The resulting restructuring is contributing not only to the intensification but also to the deregulation of work. The deregulation of work is used by organizations to appropriate employees' abilities and resources and to satisfy their needs to bring their subjectivity to work. It is, however, an open question how these intertwined trends will further develop and whether they will benefit or counteract employee well-being, motivation and personal growth in the long run (for a discussion of tendencies towards a stronger standardization of specific tasks, see Flecker et al., Chap. 2).

Another open question is how the intensification, deregulation and subjectification of work will affect employees' social embeddedness at work. As Jahoda (1981) pointed out in her seminal work on the latent functions of employment, providing workers with social contacts is an important feature of gainful employment. With the deregulation and subjectification of work, establishing modes of collaboration (sometimes among geographically scattered employees working in different time zones) becomes an individual task. Workers are obliged to organize their collaborations themselves. However, building and sustaining workable relationships becomes an inherently contradictory endeavor under conditions of intensified workloads and increased competition (Holtgrewe 2014). When work intensifies, employees may have fewer interactions with co-workers, as everyone is preoccupied with meeting his/her own objectives. Therefore, less time is devoted to actually giving support to others, which might be an important prerequisite for strong relationships at work.

In addition to the intensification of work, the almost universal trend toward using competition as a management strategy may compromise employees' social embeddedness at work. As various authors stress, competition has not only gained importance, but has also changed its character. Ilinitch et al. (1996: 211) observed a paradigmatic shift in the "rules of competition and the way the game is played" from "static competition" to "hypercompetition". Restructuring activities over the past two decades have been guided by strategies to create competition and rivalry across individuals and teams, which promise to keep organizations and their members "fit" and "slim", encourage creativity and innovation, and deliver results suited to customers' demands (Weiskopf and Loacker 2006). The new form of competition, says Bauman (2001: 123), "penetrates the obsessively dieting and slimming company of light modernity from top to bottom". Businesses "are constantly introducing an inexorable rivalry presented as healthy competition (…) that sets individuals against one another" (Deleuze 1995: 179). Yet nurturing competition and rivalry may destroy social support and social relationships at work or at least confront employees with contradictory demands: building functioning relationships and collaborating with co-workers on the one hand, and competing with them for resources and reputation within the organization on the other hand, are difficult to combine.

As shown in the chapters of this book, the trends towards intensification, deregulation and subjectification affect many, if not all, jobs. But it was also shown that the effects of these developments on employees are ambivalent. While in many cases negative effects are to be expected, under certain circumstances positive effects may also be observable. These ambivalent outcomes call for a theoretical framework allowing us to predict certain effects and guarantee a high quality of working life. As was shown in Chap. 7, Ryan and Deci's self-determination theory (SDT; e.g., Ryan and Deci 2000) could serve as such a framework. This theory postulates three basic needs: the need for autonomy, relatedness, and competence. A fourth need, the need for structure and safety, was added later by some authors (e.g., Kasser 2009). According to SDT, personal need satisfaction is an indicator of the fulfillment of these basic needs. And if basic needs at work are fulfilled, positive effects on quality of working life indicators can be expected. Thus, an evaluation of the fulfillment of basic needs in work settings could help to design jobs affected by intensification, deregulation, and subjectification with the goal of a good quality of working life.

9.2 The Reorganization of Work and "New" Dividing Lines

With the deregulation and subjectification of work, responsibility is shifted from management to individual workers. Organizations no longer provide clear (sometimes restrictive though often helpful) guidelines for when, where and how work is to be conducted or careers are to be pursued; instead, it is up to the worker to organize her/his job and career. Under such conditions, individual workers have to rely more strongly on their knowledge and skills (Warr 2013). Opportunities and developments on the labor market are therefore likely to be increasingly contingent upon employees' individual resources. Whether or not one possesses the skills necessary to actively shape one's career, optimize work and private life and regulate one's work successfully seems to be a new dividing line among employees. Having such personal resources at one's disposal might increase the likelihood of successfully entering the labor market and remaining employed. In addition to well-established resources, such as educational qualifications, social networks or economic capital, that help provide access to good jobs that promote health and motivation, individual knowledge and skills seem to represent an important form of cultural capital (see Bourdieu 1986, for the various forms of capital) that influences entrance into and success on the labor market.

According to Bourdieu (1986), cultural capital (especially in its embodied state as embodied competencies and dispositions or knowledge and skills) has to be acquired by individuals (via the expenditure of time and energy) and is an integral part of the person that cannot be transmitted instantaneously to another person. What type of cultural capital (i.e., what knowledge and skills) is valued depends on the specific context. Individual occupations, for example, may be seen as distinct arenas of social space, which Bourdieu describes as "fields" with a unique, taken-for-granted logic of practice that defines possible and acceptable behavior. Knowing the logic of practice and acting accordingly, i.e., having a "feel for the game" (Bourdieu 1990: 66), is an important prerequisite for successful conduct in a given profession. Yet the rules that define the proper functioning of the actors in the field are not explicit or codified. "Instead, they are largely implicit and partially internalised, shaping individuals' thoughts and actions to profit from or succeed within the field according to the capitals valued by it" (Watson 2013: 414). Employees whose cultural capital, acquired knowledge and skills appropriate to and legitimated by their professional context, is most closely aligned to the pervading culture of the field are likely to hold more attractive positions. In addition, their trajectories on the labor market are likely to involve greater success, especially if they are able to adapt their cultural capital (where necessary) and accrue or extend important forms of capital, while the trajectories of workers who do not possess the appropriate cultural capital or are not able to acquire it in the course of their work biography are more likely to remain in place or to lose ground (Watson 2013).

Despite the importance of competencies, knowledge and skills, that is, cultural capital, for advancement on the labor market, it is important to note that this form of capital is not independent of economic or social capital. Instead, economic capital influences the acquisition of cultural capital by determining the amount of time people can devote to acquiring skills and knowledge free of economic necessity. Social capital, on the other hand, generates opportunities for people to convert cultural capital into powerful and attractive professional positions (Bourdieu 1986). Hence, it is important to assess how the various forms of capital interact in building dividing lines among employees at a time when work is increasingly intensified, deregulated and subjectified.

9.3 Upcoming Trends in Industrial and Service Work

Apart from the above mentioned changes in the organization of work, there are a few upcoming trends observable which may further transform work and employment. As pointed out by Flecker, Fibich and Kraemer in Chap. 2, informatization and digitalization are currently important drivers of changes in work and employment. Technological advances, such as cyber-physical systems that enable communication between humans, machines and products or the internet of things and services, are likely to transform work and employment. Although it is not yet entirely clear how these developments will actually change work and employment, some potential scenarios have already been discussed in the literature upon which we will reflect in this final section. On the one hand, technological advancements might further the digitalization and virtualization of work, which will put specific demands on employees and their skills; on the other hand, they may continue to facilitate the relocation of work.

9.3.1 The Digitalization and Virtualization of Work

The debate about changes in work and employment due to technological advancements, such as the development of cyber-physical systems, is often associated with the term Industry 4.0 – a least in German-speaking countries. Protagonists of the discourse on Industry 4.0 suggest that current technological advancements will bring about a new industrial revolution, similarly to the introduction of the steam engine, the assembly line and the invention of the microchip. However, scenarios for how Industry 4.0 or cyber-physical systems will transform manufacturing industries have hitherto primarily focused on the potential of these new technologies to stimulate economic growth and increase companies' competitive advantages (Pfeiffer 2015), mostly leaving aside how they might transform work and employment conditions (see Dörre 2015, for a critical outlook on the social challenges associated with Industry 4.0).

Dombrowski and Wagner (2014), for example, assume that the integration of cyber-physical systems in the manufacturing process will increase the importance of human-machine interfaces and change employees' job profiles and required competencies. In their view, executive production tasks will decrease, while the scope of regulatory activities and error-correcting, troubleshooting and problem-solving tasks as well as interdisciplinary collaborations will increase. Hence, demands for technical or professional skills should decline and demands for comprehensive process thinking and self-organized work should increase. Independent thinking, planning and decision-making in terms of managing work tasks will become crucial competencies. In a similar vein, Brettel et al. (2014) assume that with the trend toward individualized production, control will be redirected to individual workers on the shop floor in order to ensure quick decision-making and process adaptations. These perspectives on changing demands and skill requirements closely resemble current accounts of the deregulation and subjectification of work, where heightened decision-making demands and increasing responsibility for individual workers are observed (Allvin et al. 2011; Pongratz and Voß 2003). Dombrowski and Wagner also stress that strains affecting employees may change with the shift in job profiles towards increased mental requirements. Physical strains might become less important, whereas mental strains might intensify.

While these scenarios point toward a change in skills and the potential of Industry 4.0 to promote work-based learning and upskilling (Schuh et al. 2015), other authors (e.g., Kinkel et al. 2008) warm of a deskilling and an increasing polarization of qualifications. According to the latter perspective, simple industrial work will not be abolished due to automatization, but will remain a central aspect of industrial work, as will complex, highly skilled tasks. However, jobs at the moderate skill level will erode, contributing to a polarization of competencies and skill levels.

In addition to changes in demands and skills, work arrangements which have so far primarily been realized in knowledge-intensive sectors might spread to industrial work as a result of its digitalization. Temporally and spatially flexible forms of project work may gain importance in industrial work and may contribute to the blurring of boundaries between work and non-work lives among industrial workers. Hence, we assume that the trends toward the deregulation and subjectification of work will persist or even accelerate with upcoming transformations.

9.3.2 The Relocation of Work

Additional trends highlighted within the debate on boundaryless work are outsourcing and offshoring. From the 1990s onwards, simple processes and pieces of products have been outsourced or offshored. While knowledge-intensive work such as R&D or innovation initially remained in the originating countries of multinational companies, these tasks are no longer immune to offshoring. With the gain in expertise among subcontractors in countries such as Eastern Europe, India or Russia, companies are increasingly outsourcing complex tasks to these cheaper locations (Holtgrewe 2014).

Another current trend within the relocation of work constitutes crowdsourcing employment. The term crowdsourcing was coined by Howe (2006) to define the outsourcing of work to a large group of people through an open call made possible through advances in technology and individual access to personal computers, smart phones and the internet (Barnes et al. 2013: 2). This may include the outsourcing of simple (e.g., copy editing) or complex tasks (e.g., designing graphics). Crowdsourcing for paid work has grown considerably over the last decade. In 2009, it was estimated that over one million workers engaged in this form of employment (Frei 2009 cited in Barnes et al. 2013: 20). Crowdsourcing work is particularly interesting as it raises questions about working conditions as well as skill and career development.

For individuals, participating in crowdsourcing can be both advantageous and disadvantageous (Barnes et al. 2013). This kind of work offers opportunities for flexible working. As employees find work by responding to online calls, they can choose what type of work they want to take on as well as when, where and how to work. Thus, crowdsourcing work offers high levels of decision latitude and flexibility and can thereby ease the balancing of work and non-work responsibilities. Some individuals may deliberately choose this type of employment because they desire autonomy and self-employment, yet others may be forced into crowdsourcing employment due to an inability to find appropriate work as an employee with a fixed contract. For those wishing to work part-time and/or manage their own workload, the completion of small tasks may be appropriate to their circumstances and/or expertise. However, others may have few or no alternative choices but to take whatever work they can, even if this involves working on tight deadlines and for low wages (Barnes et al. 2013). With regard to career opportunities, there are still some open questions. The current literature does not consider potentially negative effects for individuals and their employability. In particular, research assessing the longer-term trajectories and career prospects of crowdsourcing workers is absent. Yet it is obvious that this kind of work requires employees to take responsibility for their skill and career development and regulate their work individually. Crowdsourcing work is therefore an extreme example of current developments toward the deregulation and subjectification of work. As crowdsourcing work also bears the risk of exploiting employees' resources by demanding that they work on unrealistically tight deadlines, it might also reflect the trend towards the intensification of work.

As we have pointed out in this chapter, current changes in terms of the intensification, deregulation and subjectification of work represent broader patterns of transformation that are likely to affect employees in a wide range of sectors. These trends have increased the importance of personal competencies and skills, such as regulating and coordinating one's work. The individual worker and her/his abilities and skills (i.e., her/his cultural capital) have therefore also gained in importance. Upcoming trends in industrial and service work, such as the implementation of cyber-physical systems in manufacturing or crowdsourcing in service work are likely to prolong and intensify the trends toward intensification, deregulation and subjectification.

References

Allvin, M., Aronsson, G., Hagström, T., Johansson, G., & Lundberg, U. (2011). *Work without boundaries. Psychological perspectives on the new working life*. Chichester, UK: Wiley-Blackwell.

Barnes, S.-A., de Hoyos, M., Baldauf, B., Behle, H., & Green, A. (2013). *Review of state of the art and mapping: Crowdemploy*. Coventry, UK: Warwick Institute for Employment Research.

Baruch, Y., & Vardi, Y. (2015). A fresh look at the dark side of contemporary careers: Toward a realistic discourse. *British Journal of Management, 27,* 355–372. doi:10.1111/1467-8551.12107.

Bauman, Z. (2001). *Liquid modernity*. Cambridge: Polity Press.

Bourdieu, P. (1986). The forms of capital. In J. Richardson (Ed.), *Handbook of theory and research for the sociology of education* (pp. 241–258). New York: Greenwood.

Bourdieu, P. (1990). *The logic of practice*. Cambridge: Polity Press.

Brettel, M., Friederichsen, N., Keller, M., & Rosenberg, M. (2014). How virtualization, decentralization and network building change the manufacturing landscape: An Industry 4.0 perspective. *International Journal of Mechanical, Industrial Science and Engineering, 8*(1), 37–44.

Brewer, A. M. (2000). Work design for flexible work scheduling: Barriers and gender implications. *Gender, Work and Organization, 7,* 33–44.

Cascio, W. F. (2003). Changes in workers, work, and organizations. In W. C. Borman, D. R. Ilgen, & R. J. Klimoski (Eds.), *Industrial and Organizational Psychology* (Vol. 12, pp. 401–422). New York: John Wiley & Sons.

Deleuze, G. (1995). *Negotiations. 1972–1990.* New York: Columbia University Press.

Dombrowski, U., & Wagner, T. (2014). Mental strain as field of action in the 4th industrial revolution. *Procedia CIRP, 17,* 100–105.

Dörre, K. (2015). Digitalisierung – neue Prosperität oder Vertiefung gesellschaftlicher Spaltungen? [Digitalization—New prosperity or deepening of societal cleavages?] In H. Hirsch-Kreinsen, P. Ittermann, & J. Niehaus (Eds.), *Digitalisierung industrieller Arbeit. Die Vision Industrie 4.0 und ihre sozialen Herausforderungen* (pp. 269–284). Baden-Baden: Nomos.

Garhammer, M. (2002). Pace of life and enjoyment of life. *Journal of Happiness Studies, 3,* 217–256.

Holtgrewe, U. (2014). New new technologies: The future and the present of work in information and communication technology. *New Technology, Work and Employment, 29,* 9–24.

Howe, J. (2006). The rise of crowdsourcing. *Wired, 14*(6).

Ilinitch, A. Y., D'Aveni, R., & Levin, A. (1996). New organizational forms and strategies for managing in hypercompetitive environments. *Organization Science, 7,* 211–20.

Jahoda, M. (1981). Work, employment, and unemployment: Values, theories, and approaches in social research. *American Psychologist, 36,* 184–191.

Kasser, T. (2009). Psychological need satisfaction, personal well-being, and ecological sustainability. *Ecopsychology, 1,* 175–180. doi:10.1089/eco.2009.0025.

Kelliher, C., & Anderson, D. (2010). Doing more with less? Flexible work practices and the intensification of work. *Human Relations, 63,* 83–106.

Kinkel, S., Friedewald, M., Hüsing, B., Lay, G., & Lindner, R. (2008). Arbeiten in der Zukunft: Strukturen und Trends der Industriearbeit. [Work in the future: Structures and trends of industrial work]. Berlin: Sigma.

Pfeiffer, S. (2015). Industrie 4.0 und die Digitalisierung der Produktion – Hype oder Megatrend? [Industry 4.0 and the digitalization of production—Hype or megatrend?] *Aus Politik und Zeitgeschichte, 31–32,* 6–12. Retrieved from http://www.bpb.de/apuz/209955/industrie-4-0-und-die-digitalisierung-der-produktion?p=all#footnode3-3

Pongratz, H. J., & Voß, G. G. (2003). From employee to 'entreployee'. Towards a 'self-entrepreneurial' work force? *Concepts and Transformations, 8,* 239–254.

Ryan, R. M., & Deci, E. L. (2000). Self-determination theory and the facilitation of intrinsic motivation, social development, and well-being. *American Psychologist, 55,* 68–78. doi:10.1037/0003.066X.55.1.68.

Schuh, G., Gartzem, T., Rodenhauser, T., & Marks, A. (2015). Promoting work-based learning through industry 4.0. *Procedia CIRP, 32,* 82–87.

Tripoli, A. M. (1998). Planning and allocating: Strategies for managing priorities in complex jobs. *European Journal of Work and Organizational Psychology, 7,* 455–476.

Warr, P. (2013). Jobs and job-holders: Two sources of happiness and unhappiness. In S. A. David, I. Boniwell, & A. Conley Ayers (Eds.), *The Oxford Handbook of Happiness* (pp. 733–750). Oxford: Oxford University Press.

Watson, J. (2013). Profitable portfolios: Capital that counts in higher education. *British Journal of Sociology of Education, 34,* 412–430. doi:10.1080/01425692.2012.710005.

Weiskopf, R., & Loacker, B. (2006). A snake's coils are even more intricate than a mole's burrow: Individualization and subjectification in post-disciplinary regimes of work. *Management Revue, 17,* 395–419.

Wood, L. A. (2011). *The changing nature of jobs: A meta-analysis examining changes in job characteristics over time.* Master thesis, University of Georgia, Georgia, USA. Retrieved from http://dbs.galib.uga.edu/cgi-bin/getd.cgi?userid=galileo&serverno=8&instcode=publ&_cc=1

Index

C. Korunka and B. Kubicek (eds.), *Job Demands in a Changing World of Work*,
DOI 10.1007/978-3-319-54678-0

MIX
Papier aus verantwortungsvollen Quellen
Paper from responsible sources
FSC® C105338

If you have any concerns about our products,
you can contact us on
ProductSafety@springernature.com

In case Publisher is established outside the EU,
the EU authorized representative is:
Springer Nature Customer Service Center GmbH
Europaplatz 3, 69115 Heidelberg, Germany

Printed by Libri Plureos GmbH
in Hamburg, Germany